PLANTING WETLANDS + DAMS

A Practical Guide to Wetland Design, Construction + Propagation

NICK **ROMANOWSKI**

Nick Romanowski is a wetland consultant with a background in zoology and considerable practical experience in recreating the habitat needs of aquatic animals. His nursery, Dragonfly Aquatics, was the first to specialise in Australian water and wetland plants. The nursery is also home to visiting waterbirds and a resident population of many native animals. Nick is the author of the companion volume to this book, *Aquatic and Wetland Plants: A Field Guide for Non-tropical Australia*, and *Water Garden Plants and Animals: The Complete Guide for all Australia*, both published by UNSW Press. He is also the author of several other books, including *Water and Wetland Plants for Southern Australia* and *Farming in Ponds and Dams: An Introduction to Freshwater Aquaculture in Australia.* His articles on aquatic plants and animals have been widely published in journals.

A UNSW Press book

Published by
University of New South Wales Press Ltd
UNSW Sydney NSW 2052
AUSTRALIA
www.unswpress.com.au

First published in 1998
Reprinted 2000

National Library of Australia
Cataloguing-in-Publication entry:

Romanowski, Nick, 1954– .
Planting wetlands and dams: a practical guide to wetland design, construction and propagation.

ISBN 086840 608 2.

1. Wetland landscape design—Australia. 2. Dams—Australia—Design and construction. 3. Wetland plants—Australia. I. Title.

577.680994

Printer Hyde Park Press

Cover The degrees to which different species adapt to local conditions can vary considerably. These three plants in a coastal wetland are good examples. *Villarsia reniformis,* with floating, kidney-shaped leaves, is confined to south-eastern Australia, but a closely related species occurs in the south-west. *Myriophyllum salsugineum,* with flower columns above water, is found in many southern areas, but some eastern and western populations have been separated for long enough to be noticeably different. The sedge *Baumea juncea* is found across much of non-tropical Australia and beyond, yet varies little across this considerable range.

CONTENTS

ACKNOWLEDGMENTS

This book could not have been written without considerable help and encouragement from many people over a long period of time. I would particularly like to thank Surrey Jacobs for his patient advice over the years, and Helen Aston, not only for more recent assistance but also for triggering my initial interest when *Aquatic Plants of Australia* first came out in 1973. Particular thanks also to Ian Bayly and Bill Williams, who guided my early steps in the systematic study of freshwater animals and their environments.

People who have helped with comments and advice of many kinds, from the behaviour of clays to plant propagation and animal identification, include Tony Brindley, Norman Deno, Tim Entwisle, Rodger Elliot, Howard Gill, Pierre Horwitz, Paddy Lightfoot, Howie Marshall, Ian McCann, John McCubbing, Graham Milledge, David Morgan, Bob Myers, Walter Pagels, Andrew Paget, Anna and Pino Pistillo, Mike Schulz, Nick van Roosendael, Robin Wilson and Geoff Winning. My understanding of wetland plants has been considerably enhanced through discussions and correspondence with a number of botanists including David Albrecht, Helen Aston, Ian Clarke, Tim Entwisle, Helen Hewson, Surrey Jacobs, Tony Orchard, Bob Ornduff, Jim Ross, Roger Spencer, Val Stajsic, Neville Walsh, Sabby Wilkins (Papassotiriou) and Karen Wilson.

Parts of this book have appeared in a preliminary form in the journals *Indigenotes*, *Land for Wildlife News* and *Wetland Ways*. I would like to express appreciation to the respective journal editors Lincoln Kern, Stephen Platt and John Jago for the feedback this has generated. Although all of the illustrations and most of the photography included are my own, Surrey Jacobs has also generously allowed me to use photos of Lake Pelion and *Monochoria australasica* from his own slide library.

My wife, Jan Ratcliff, has taught me a great deal about writing and organisation over the years, acting also as my initial editor. This book could not have been written without her encouragement and advice. It is dedicated to our two Mudeyes (slang for aquatic Dragonfly larvae) Katya and Talya Romanowski, who have taught me more about mud than anyone could reasonably need to know.

INTRODUCTION

This book brings together some of the plant-related facets of my life-long interest in wetlands, from an early interest in aquatic animals and their habitats to my work with plants over the past decade. As an incidental spin-off, this led me to set up the first nursery to specialise in Australian water and wetland plants, Dragonfly Aquatics. Much of the new practical information on plants included here has come from this, and from the many contacts I have made in that time.

My original intention was to produce a more complete guide to the practical aspects of wetland creation, restoration and management, but after more than 2 years of preparation and writing it became obvious that planned sections on water treatment and on the habitat requirements of wetland animals would need a separate book each if they were to be adequately covered! Over the same time it also became obvious that the most pressing void in wetland information today is in areas such as plant identification, propagation and planting out, and the role of plants as habitat or in water treatment.

This book is one of two volumes intended to fill most such gaps (the companion volume, *Aquatic and Wetland Plants: A Field Guide for Non-tropical Australia*, is a non-technical guide to the identification of about 340 species). I hope it will help to generate a greater awareness of our neglected and often endangered wetlands, of their value, and of the steps we can and should be taking to restore, replace and replant wherever we can.

1 WETLANDS

Living organisms don't just need water. By weight, every living organism *is* mostly water. Nearly all water on the surface of this planet is in the oceans, which cover most of the thin surface layer of rock we live upon. The heat of the Sun evaporates water from the sea, forming the rain-bearing clouds on which the lives of all land-dwellers depend. Rain brings water to the land, and the water flows mostly downwards as it returns to the oceans, picking up minerals and other nutrients and forming wetlands of many kinds as it goes.

It isn't easy to define just what a wetland is, and there are different definitions by hydrologists, legislators, engineers, biologists and ecologists. In the USA an entire job category entitled 'wetland delineators' is required to work out where wetlands begin and end! Some definitions are legal and some are practical, but the best ones are common-sense: wetlands are places where water forms pools or flows that last long enough for plants and animals to base a significant part of their year or lifespan around their existence. Some wetlands are permanent, forming bodies of water that rarely disappear; others may flow just beneath the surface of the soil, rising and falling with seasonal rainfall; still others may hold water for only a few months.

Plants and animals found in the many types of wetlands are often adapted to extremes of flood and drought, but all need water for at least a part of their life cycle. Some may need water to quench their thirst, while others live in it for most of their lives and only come out when they must breed. Of course, many must have water for every stage of their existence. Many wetland plants and animals are important even to surrounding terrestrial ecosystems, and the wetlands themselves recapture many nutrients and much soil that would otherwise wash out to sea and be lost to the land.

Some theories suggest that wetlands have been a driving force for part of human evolution, and these would go a long way to explaining some unusual aspects of our biology and perhaps lifestyles as well. We are still drawn to water, and an estimated three-quarters of all human recreational activities take place in, on or near it. Even an evening stroll is more pleasant near a creek or lake! The result of this urge to be near water is that towns near large areas of wetlands draw tourists on a much larger scale than those without.

The role of wetlands in water treatment is also being increasingly appreciated, even for such potent problems as human sewage. On private land they can act as seasonal grazing and drought-proofing, and add to the selling value of the property. There are other obvious scientific, educational and aesthetic values that can't be costed in dollar terms. And even these things may be minor compared with the long-term values of wetlands, which include groundwater recharge. We still know relatively little about the role wetlands play in maintaining the quality of underground flows, yet it is already obvious that much of the useful underground water in Australia will have been used up (as it already is in the USA) or else will be fouled beyond use within the next few decades if we continue to treat wetlands as waste dumps.

On our increasingly polluted planet, Australia can claim to be among the last remaining sources of relatively pure, natural products, most of which depend directly or indirectly on the quality of our wetlands. Yet the sad fact is that on this driest of liveable continents we have put an incredible amount of effort into draining and destroying this precious resource. For example, we have lost at least one-third of wetlands in Victoria, three-quarters around Sydney and on the Swan coastal plain around Perth, and nine-tenths in south-eastern South Australia.

The results are already obvious. Stormwaters drain off to sea more rapidly than ever before, and carry more soil and nutrients than ever before as well. Floods are more serious because most watercourses have been altered so much that flow rates and spillover are less controlled by the natural processes that used to regulate them in the past. The loss of our wetlands also endangers a substantial proportion of our fauna and flora, silts and pollutes rivers and underground waters, and makes Australia a bit drier, greyer and quite a bit less beautiful. The damage can and must be undone if we are to salvage our fragile ecosystems instead of continuing to follow the crude and ill-informed practices brought with the first European settlers.

• TYPES OF WETLANDS

There is a rich diversity of wetlands in Australia, from coastal-cliff waterfalls (photo 1) to swamps, rivers and springs, and from deep lakes to roadside ditches that are dry for most of the year. They are classified in many different ways by the various groups of people who study them, based on the way the water moves through the system, plant types, animals habitats, or other aspects. Such classification systems are usually designed as a framework for particular ways of studying wetlands, and aren't necessary for a more general appreciation of their diversity.

The main 'types' of wetland are briefly described here, with a particular emphasis on those that are likely to occur on private lands, council reserves, and other places where their management and well-being can be helped along considerably by private individuals and groups.

DAMPLANDS AND SUMPLANDS

Damplands aren't usually thought of as wetlands because much of their water is below the soil, sinking deep during drier seasons and rising to saturate the surface during wet seasons. (The level of this underground water is called the water table.) Damplands are often acidic and poor in nutrients, yet many support a diverse range of flowering plants (photo 2) including melaleucas, tea trees, heaths, sedges (from tiny creeping species to giant saw sedges), and other grass-like plants from restiads and rushes to yellow-eyes. The shortage of some nutrients here (especially useable nitrogen) has been the cause of much specialisation among plants, even to extremes such as carnivorous plants that obtain their needs by trapping insects.

Plant matter doesn't break down rapidly in such nutrient-poor, waterlogged soils, and may build up very gradually as peat. Peat will hold large amounts of water for long periods of time, releasing it slowly into wetlands downstream even after the water table has sunk far below the dampland. For this reason, damplands have a significant effect both on reducing flooding during periods of heavy rain and in maintaining the flow of rivers and creeks during dry periods.

Where the water table is closer to the surface, it may rise above the soil level during the wet season, filling all depressions with water and leaving even the higher ground sodden. This much wetter type of area caused by rising water (rather than by flooding) is called sumpland.

EPHEMERAL WETLANDS

Ephemeral wetlands fill directly from rain and last only for a few weeks or months. The plants in them may die back to corms, rootsocks or soil seed banks until better conditions return, while the animals will move elsewhere or produce drought-tolerant eggs or cysts (in many species the breeding animals don't live more than one wet season). Other animals (particularly in arid inland areas) may burrow into the ground where they may survive for years while waiting for the next rains. While such conditions seem harsh to us, they have great advantages for suitably adapted organisms. Ephemeral wetlands may literally swarm with life for a time, and are free of predators that cannot survive dry spells.

Spore, cyst and seed stages of plants and animals from ephemeral wetlands are often very small and lightweight as they are designed to be carried between wetlands by wind, waterbirds or possibly even insects. Arriving in a suitable place they will thrive, and it isn't generally appreciated how many suitable places are found on private lands. Nor is it appreciated how little work or expense is needed to open up new potential habitat for them; even roadside ditches (photo 3) may harbour numerous planktonic species. Each new ephemeral wetland is also a further extension to the feeding grounds of many waterbirds and insects, which can visit during the wet season and leave when conditions are no longer to their liking.

FERN GULLIES AND SPRINGS

This assortment is not a part of anyone's wetland classification, but is my own way of grouping wetlands in which water moves fairly rapidly underground, sometimes even surfacing. There is a practical reason for this grouping; moving water often carries a reasonable amount of oxygen, even underground, so that plants that would drown in stagnant conditions will thrive on slopes even though their soil is completely waterlogged. This is probably why some ferns grow well on waterlogged sloping sites but are only seen on higher ground around swamps where the water is not moving.

The types of plants found around springs vary considerably, depending partly on the nature of the water as this can be anything from soft and cold to alkaline and hot. Completely different types of spring may even surface close together in some areas. The plants of fern gullies may also vary, but more because of variations in light than in water quality, which is usually relatively soft. Tree ferns and associated species will dominate in shaded gullies, while on brightly lit and permanently wet sites the dominant species are more likely to be coral ferns, water ferns and king ferns (photo 4), with a canopy of melaleucas above.

MARSHES AND SWAMPS

Extensive areas of still, reasonably permanent wetlands are generally called marshes or swamps. Their water levels may rise and fall to some degree between wet and dry seasons, but there is often some water remaining near the surface, at least in most years. Marshes and swamps can spread over great distances in the wet season, and are of considerable importance as habitat in many ways. Although accurate figures are not available for much of Australia, there is no doubt that a large proportion of the remaining wetland of these types lies on private land.

No simple generalisations can be made about marshes and swamps; they are remarkably diverse, and even nearby swamps can be dramatically different in the plants and animals that dominate them. In part, this could be because the first two or three plants to establish in a swamp may be able to prevent later arrivals from establishing themselves properly. In turn, many animal species may only be found where certain plants are already established, although this is better documented for insects rather than for larger animals.

Marshes and swamps are among the hardest hit of wetlands in post-1788 Australia. They have been drained for agriculture, grazed by livestock, poisoned to control mosquitoes and mangled by feral pigs. They are also among the types of wetlands that potentially can benefit most from good management and protection by private individuals. Their great diversity is well worth preserving.

RIVERS, STREAMS AND CREEKS

Rivers, creeks and streams should need no defining, although they can vary considerably in type from cold, fast-flowing mountain brooks to sluggish, warm and often muddy inland rivers. Theoretically, the management of such moving waters usually falls under various types of government control, even including some of the surrounding land. In practice, riverside vegetation is often managed or mismanaged by landowners and local government.

The worst forms of abuse are grazing and clearing, which expose riverbanks to erosion (photo 6), especially during flood periods, and contribute significantly to increasing siltation and loss of water quality downstream. This also reduces the habitat diversity and food supply of many animals adapted to waters running off cleaner and healthier catchments. Other abuses are less obvious, and include the planting of willows along river banks in preference to indigenous vegetation. Although many groups are doing a good job of carrying out and encouraging re-vegetation, the health of rivers, creeks and streams will ultimately only be improved by action through their whole catchment. This is likely to happen only if state and federal governments finally understand that we are presently flushing away a large part of our natural wealth, and take real action to protect riversides instead of just pretending that they are doing so.

FLOOD PLAINS AND BILLABONGS

Rivers flowing through flood plains (particularly near the coast, and through much of the Murray-Darling system) tend to meander as they cut their ways across. These plains of silt are deposited during periods of flood and can be very extensive; after floods the various depressions in them fill to form swamps and pools. The most familiar name for these in Australia is 'billabong', referring to a river bend that has been isolated from the main river flow when a new channel is cut (photo 5). As the openings to the new river channel silt up, the new billabong becomes a still waterbody that usually depends on floodwaters to refill it.

Billabongs and other flood plain swamps are fascinating worlds in their own right, both a part of the river and separate from it. Their fauna and flora can be very rich, and may include species that aren't often seen in other types of wetland. With the increasing control and flow regulation of many rivers, particularly the inland ones, it is likely that the formation of new billabongs is becoming a rare event compared to what would have been happening in the past. At the same time, the disruption of natural flood regimes is killing off the species diversity of the old ones.

Until water for wetlands is regarded as a basic right (rather than being mostly directed to agriculture), billabongs and flood plain swamps are only likely to survive in a reasonably natural state with enlightened management, and then only if seasonal flooding can be arranged. It is possible to create artificial billabongs on flood plains, and this would be worth developing in the future as a way of maintaining adequate habitat for the more specialised animals and plants that depend on such waters.

LAKES

Lakes vary considerably as they are formed in many ways: tarns scoured out by glaciers in the high country (photo 7); volcanic craters (themselves very variable); valleys dammed by lava flows or landslides; and hollows behind sand dunes that accumulate organic matter until they hold water. Their waters vary from nearly as pure as distilled water through to acidic and peaty, clear or soupy with microscopic life, and brackish to more salty than the sea. Living conditions in any body of water large enough to be called a lake will be very different to those in other types of wetland.

Lakes are subject to strong winds and wave action, so the vegetation around their fringes is often species-poor, being limited to plants that can tolerate relatively rough conditions. Conditions like these are also seen in some artificial wetlands that are constructed to be lake-like, making it virtually impossible to establish vegetation on the exposed and wave-lashed shores! Lake-dwelling animals may also need to be tolerant of other extremes; for example, in deeper lakes the upper and lower layers may have quite different conditions for most of the time, only mixing at the two times of the year when the surface and bottom layers equalise in temperature.

UNDERGROUND WETLANDS

There is far more water hidden underground than is revealed by surface wetlands, although most of it is retained in soils and sand rather than forming pools. However, there are also extensive cave systems with pools and streams, particularly in limestone areas, and underground waters associated with rivers and springs. Although various types of animals, especially crustaceans, are adapted to life in these hidden places, plants are virtually absent (unless you consider fungi to be plants). Virtually nothing is known about the management of subterranean waters, but simply maintaining a diversity of well-planted wetlands up above is likely to help keep them healthy.

SALT MARSHES

Saline wetlands have a very distinctive flora, much of which is shared between inland areas and the salt marshes behind mangrove belts on the coast. At first glance, salt marshes may seem rather bleak because of the absence of trees, which can't establish because salinity levels during periods of low water flow are far beyond the tolerances of most plants. A closer look will reveal some fascinating plants adapted to the extreme conditions prevailing, often arranged in zones that reflect their tolerance of salt or flooding with freshwater, and to varying degrees their tolerance of drought in some seasons.

Little is known about revegetation and restoration of salt marsh species because there hasn't been much need for this as the lands they occupy have rarely been considered useful for agricultural purposes. However, this situation is rapidly changing as an increasing number of coastal salt marshes are destroyed along with mangroves to make room for development. The planting or revegetation of saline soils will also become increasingly important as salinity problems associated with the past two centuries of unintelligent land management continue to take their toll.

ESTUARIES AND MANGROVES

By most definitions, wetlands end somewhere along the sea's edge, and where the transition is fairly gradual there will often be trees or shrubs called mangroves (photo 8), their feet placed in the zone of lowest tides where the seagrasses begin. Mangrove species are largely found in estuaries, where seawater and freshwater mix. The sea pushes in at high tide, but later the rivers and creeks push the saltwater back downstream as the tide falls. The mixing of freshwater and saltwater varies with flood and tide to give diverse salinities at different times and places through the estuary, although in smooth conditions the less dense freshwaters may ride out to sea above the denser seawater below.

These areas are perhaps the most important feeding and nursery grounds for a diverse range of fishes and other marine animals, but they are rapidly being destroyed by developers anxious to make their fortunes from the many people who love the sea and wish to live near it. It is a tragedy that by paying for the conversion of mangrove and estuary into housing estate, they are helping to destroy much of what they would like to preserve: clean water, good fishing and the beauty of the coast.

FARM DAMS AND OTHER ARTIFICIAL WETLANDS

Farm dams and other artificial wetlands are a considerable addition to our depleted natural wetlands. They represent a major increase in both area and types of habitat, and with suitable modification most can be made still more useful as an extension of natural wetlands. There are probably close to half a million farm dams in Australia, and any flight over drier areas will show that these are almost the only kind of open water available over immense landscapes. Many are in areas where there would normally be little or no permanent water, or where the artesian springs that once were the natural waters are now disappearing as a result of excessive pumping from bores.

Other types of artificial wetlands have only been constructed in any numbers in the past decade or so, so there are still relatively few of these. They represent not only an opportunity to try to recreate many types of natural waters, but also to study the dynamics of wetland ecology in ways that are simply not possible in more passive types of research. Dams and artificial wetlands may also have another role to play. Those that can be kept free of introduced weeds and vermin could ultimately be used as refuges for species that can't compete against the new-comers, while such artificial waters may be the only real chance of survival some endangered species have left (photo 9).

It isn't possible to duplicate the ecology of a natural area in a newly created wetland. Indeed, it is not possible to recreate some types of wetland at all with our present knowledge. However, a properly designed wetland will have the potential to mature into something very close to the real thing, and may be indistinguishable within a decade or so. Created wetlands are rarely intended to just imitate nature, though, and many other values including water treatment, aesthetics and ease of access should also be considered in their design. For some purposes, there may even be advantages to the creation of visibly artificial waterbodies rather than recreating the problems facing natural wetlands today.

Whatever the purpose of a created wetland, the commonsense approach to design and planting will use species that occur naturally in the area using undegraded local wetland communities as models. Such places have evolved with and are already adapted to the challenges of climate and season, and include a rich diversity of species compared with those that have been seriously affected by grazing, drainage or salinity. Of course, some plants that don't occur in the area may turn out to be even better suited to local conditions if they are included. These are called weeds.

2 PLANNING A DAM OR WETLAND

Most wetlands and many dams are created for a variety of purposes. These should be clear in the designer's mind before commencing the planning process as they are not necessarily all compatible or easily achieved in a single body of water. Perhaps the single most common purpose intended is habitat for the varied and endlessly fascinating plants and animals found around water, yet most created wetlands in Australia are treated as an exercise in landscaping. No one expects to recreate a forest ecosystem in an artificial planting, but many designers believe that they are recreating ecologies even if they have little knowledge of their component parts.

Another frequent problem with the current notions of wetland planning is that there is rarely any assessment made of whether a created wetland does what it was supposed to do. Proper planning should also encourage recording of results, whether these come out as expected or not. This will not only help in creating more predictable and easily managed wetlands later, but can even add to our understanding of the ways in which the natural systems work. This chapter outlines the general ideas behind planning, including related areas such as modification of existing dams to make them more like natural waters.

• WHAT ARE YOUR GOALS?

Planning is largely a matter of common sense, and it is usually possible to blend various goals in one wetland, although some may have to be modified. Start by listing those that are central to your purposes and those that can be adapted to fit around the others; it doesn't take more than a few minutes of thought to see where problems and incompatibilities may set in. For example, swimming (photo 10) can combine well with some types of aquaculture, but not with stock access for watering. By contrast, good fishing can be had even with some stock watering, but is not compatible with most other habitat requirements unless no hooks or nets are likely to be lost where animals can swallow them or become entangled.

Appearance is usually important in artificial wetland design, particularly maintaining enough areas where open water can be seen from the shore. This is sometimes done by using only very low-growing plants in the shallows, but the resulting effect is both artificial and has limited habitat value. It is more effective to combine a greater diversity of plant species with varied underwater planting depths to give a more natural looking shoreline.

Irrigation from a wetland is not compatible with most other goals unless the wetland area is very large and the volume of water to be taken is small. Natural wetland levels drop during the warmer months through evaporation or lowered water tables, but if you add irrigation drawdowns to this equation then many wetland species won't be able to complete their natural cycles in the reduced time available. If you want irrigation water *and* a wetland, it is best to create a water storage dam upstream/uphill of the wetland; this can be used to top up the wetland as well if required. However, a wetland associated with a dam is a different matter, and this is considered later.

One general recommendation can be made for most types of created wetland: make them as large as you can within the space you have available and the limitations of your budget. Cost is often a factor for extensive works as earthmoving is expensive, but the cost can be spread over a number of years when a series of wetlands are made. If these are to be linked, make allowance for each one to join established areas in such a way that new work doesn't suddenly drop the water level in already established areas. A further advantage of gradual extension is that you can use thinnings and seed from earlier plantings in new areas as they are constructed.

• REGULATIONS AND PERMITS

Laws and permits affecting the construction and restoration of wetlands vary considerably from state to state, and any attempt to summarise the current situation for all of Australia would require a book in itself. However, it isn't difficult to give a general idea of who to approach for information, and this will lead you through the legal problems potentially associated with work of this kind.

A new wetland is least likely to need elaborate paperwork in rural areas, or at least no more than would usually be associated with sinking a dam. The local shire or council usually makes all decisions of this kind, and you may be given approval to go ahead on the basis of a phone call. However, it is best to request confirmation in writing. Any unexpected restrictions or hitches are most likely to be discovered if the contact person has to check up on their facts before committing themselves to writing.

There will be limitations on what can be done even at such a basic level, particularly where a dam or wetland may bank water up so that it overflows onto a neighbouring property. This is obviously unacceptable unless the neighbour is keen to share in the new wetland. Permanently flowing waters can't be dammed without permission as this is usually regarded as a public asset rather than a private one, and there may also be specialised safety considerations for water-retaining embankments in such cases.

Potential effects on any kind of indigenous vegetation already present must also be considered, and in some states it is illegal to deliberately or accidentally drown other vegetation types. In any case, it should be obvious to anyone wanting to create new habitat of this kind that such damage is environmental vandalism.

For all these reasons, the future boundaries of any new wetland must be fully surveyed before plans are finalised. If no problems are found, and the water running out of the new wetland will be at least of the same quality as what is running in, then usually no other permits are needed. Of course, if the water running out of the wetland isn't improved in quality on its way through, something is seriously wrong with the design!

There will often be more complicated procedures to go through if the planned work involves modification or restoration of an existing wetland. There are good reasons for not allowing random alteration of even the most weed-infested and degraded wetland. For example, valuable remnant populations of less common plants may still be present, even if these are largely swamped by weeds and unable to grow or reproduce properly. Any alterations made should include a management plan for such remnants, giving them a chance to recover as a result of the changes.

Extending existing wetlands requires even more care and thought because this invariably shifts wetland boundaries and alters flooding regimes. All wetland plants have definite water depth preferences, so raising water levels too rapidly and too much will drown those that aren't able to adapt fast enough. A very gradual increase over many years will allow most wetland plants to shift with the new levels. However, plants that can't sprout runners may be unable to colonise areas where other species are already entrenched and would need transplanting, which is not always successful.

Most works that will affect existing vegetation are regulated by state laws rather than local ones. They are administered by whichever authority protects native fauna and flora in your state, although if you plan to introduce fishes there may be fisheries laws to comply with as well. Check all such laws through your regional state government offices. It is rarely difficult to make contact with the person best able to tell you which laws will be relevant and whether there is any other state office (usually in the same building) you may need to contact.

In more built-up and urban areas there is a much greater need for following up all laws and permits that could possibly impinge on the planned work. The reasons are obvious and may include neighbours who are likely to be closer, existing plant communities that are more likely to be under pressure from weed competition, and the safety of dam walls and embankments. Contact with local government remains the first step to finding what legal constraints there may be, including information on the possible presence of underground lines and pipes and who has put them there.

After overcoming any complexities of wetland construction and restoration, you may also need to apply for a permit to collect seed, cuttings or divisions of plants for propagation if these are to be taken from public land. Such permits are also managed by state and territory governments. Collecting from private properties can usually be done with just the owner's permission, but there may be state restrictions on which types of plant can be gathered even here.

Whatever plants are collected, there is often a requirement to keep records of what has been harvested and in what quantity. It would be better still to keep records of how these plants establish and grow as well. If possible, take photographs to document the changes every few years, although there is no legal obligation to do this anywhere. Wetland planting is, relatively speaking, still in its infancy. The more records that are kept, the more information is exchanged and the sooner the creation and restoration of wetlands will become a science that will produce useful and predictable results.

• CHOOSING SITES

The two most important factors in creating most types of wetlands are cost-effectiveness (area created per dollar spent) and natural appearance. There is no conflict between the two: the least expensive, most natural looking wetlands are on relatively flat land with some slight undulation and variation. The less earthmoving the better, so islands, peninsulas and underwater planting shelves or plateaus should follow existing contours as closely as possible. On the minus side, such sites may already have wet or boggy patches with established plants and, if any of these are unwanted species, both plants and seeds will need to be controlled before the wetland is flooded.

Some low-lying sites may also be unsuitable as homes for smaller native fishes and frogs, many of which have been considerably reduced in numbers through competition with gambusia (*Gambusia holbrooki*). This aggressive little fish was introduced to control mosquito larvae, which it doesn't do any better than native fishes. It is now very widespread, and will frequently arrive in new waters during floods. So will carp (*Cyprinus carpio*), which are very destructive towards submerged vegetation and also indirectly reduce water quality for other animals.

To keep both these fishes out, earth banks (levees) may need to be constructed around low-lying wetlands; these should be engineered to hold off even the highest waters. Levees are usually designed to cope with lesser degrees of flooding; 20 or 50 year events are commonly specified. However, even one breach of the embankment in 20–50 years can bring in unwanted fishes, which are often difficult to eradicate without complete draining of the system. If it is not possible to build up a higher levee, a more elevated site may be preferable if the wetland is intended to provide satisfactory fish or frog habitat.

Wherever the wetland is sited, it is likely to bank water up into areas that would normally have been drier. This increased flooding can kill plants that aren't adapted to wet feet, and even water's edge species won't survive an extra month or two to their average wet period. Commonly seen examples of this problem include redgum (photo 14) or melaleuca stands that have died gradually because of permanent rises in water levels, sometimes even quite small ones.

The water source for the wetland must also be considered. This is usually either run-off water that is collected in a water-retaining basin or an underground water table that rises and falls with the seasons. Wetlands designed to collect water that runs into them should be constructed in the same way as a properly made farm dam or they will leak as is the case in an estimated one in three farm dams. The area of such wetlands will partly depend on the amount of run-off water flowing in from above them, as there is no point creating a large wetland area that doesn't have enough catchment above it to fill it in an average year of rainfall.

It isn't always easy to work out how much water will be available; this depends on when and for how long the rain falls, how much water the soil will absorb before further rainfall begins to run off, and evaporation rates. If you are in any doubt that the catchment is adequate for your plans, then your plans are probably too ambitious. However, it may be possible to increase the catchment area with shallow drains fanning outwards to collect run-off water from a greater area than would normally be available.

Wetlands should not be dug into an existing water table if you aren't sure how much it will rise and fall with the seasons. This can be worked out by using an earth auger to dig a test bore. Even a hand-driven 10 cm auger can be used for this, and can be extended up to 3 metres with additional pipe sections of the same thread as you go deeper. Where the water table is found to vary regularly by a metre or more, there will be problems with establishing water plants, just as in irrigation dams.

Where underground water is surfacing from a spring, it is best to avoid disturbance as this can alter flow rates or even stop the flow altogether. In extreme cases where a spring enters below the planned water level, water may even channel back out through the spring so that the dam or wetland never fills. Some very deep artificial pools and dams are spring-fed, but usually have the spring at the side of the pool or slightly above it so that it is minimally disturbed by digging. Springs on sloping ground can also be used to fill a wetland below.

Although relatively flat or gently undulating land is the easiest and least expensive place to build substantial areas of wetland, not everyone has a site like this. The greater the slope of a site, the less likely there is to be any underground water table that can usefully be tapped into. On sloping sites the most common option is likely to be a wetland built like a dam using a water-retaining wall. Extensive wetland areas are impractical in such situations, but a series of smaller wetlands descending from one to the next may be possible, although expensive.

Where there is room for only a single wetland on a relatively steep site, it will need to be deep for its surface area and have a substantial retaining wall (in fact, it will be very much like a dam). There is nothing wrong with this assuming that you can afford it; well designed dams can be used in various ways to create a diverse range of habitats. However, the deep waters and steep sides of walls and islands in a dam are often incompatible with the types of plants seen in broad, shallow wetlands, and may be undercut along shorelines by waves and dabbling ducks.

Sites with excessively saline groundwater should be avoided. Although a few plants will grow and even thrive in very brackish water, relatively few animal species are as tolerant. This means that saline waters may build up large populations of midges and mosquitoes, the larvae of which are eaten by a wide range of predators in fresher waters. On the positive side, a moderate degree of salinity may be an advantage in dispersive clays as it will help to bind clay particles together so that retaining walls hold water more reliably.

Salinity is a more subtle threat in dams and wetlands filled by run-off, especially in drier climates. The run-off water may only be moderately saline, but the salinity in the wetland will increase in warmer seasons as evaporation concentrates the salts. If such dams and wetlands aren't adequately flushed by winter rains, their salinity will continue to climb from year to year.

• OTHER CONSIDERATIONS

Depending on the uses you intend for your wetland, you may need to plan for reserve water supplies, structures extending out over water, islands of various kinds, a varied underwater landscape, allow for overflow in flood times, remove excessive sediments from incoming water, or make provision for drawdown of water levels. Drawdown and overflow structures are discussed in the next chapter under 'Water Control'; the other aspects are considered here.

Structures that project out over water should have their footings in place before the wetland fills, which often means within a few weeks of completing the earthworks. Ramps for foot traffic (photo 15) can be constructed on established footings even after water levels rise, but for more elaborate structures, such as jetties and hides, the supporting frame or scaffolding should already be in place before flooding.

Ramps may be built over any range of water depths or wet soils, but jetties and hides should be placed so that they finish well out from the shallowest water, whether they are used for fishing, swimming, or observation. It is all too common to see a hide that doesn't extend beyond the shallows, with a tall screen of reeds or rushes in the foreground obscuring more distant views. Hides are intended to give a good view of birds and their behaviours, and although plants can be cleared out of the way regularly, this may make the nearby shallows less attractive to waterbirds. Viewing windows at the sides of the hide are a more appropriate place to observe nesting areas as plants can be allowed to spread here as they see fit.

Jetties are a very different type of observation area, often sitting lower to the water and allowing better viewing of underwater animals, including fishes. Jetties reduce the reflection of light that, from the shore, can obscure much of what is visible below the surface compared with looking more-or-less straight down into the water from atop a jetty. Fishes, in particular, are less wary near jetties because, from their point of view, the platform you are on looks like a deeply-shaded, overhanging ledge under which they can easily take cover.

Other uses for jetties include as a working area next to the water, as a place to fish, to hang aquaculture cages, and they are also a good place to hang a hammock. (photo 10). There are many other ways in which they can be adapted. For example, they can be designed so that temporary screening walls can be clipped into place to make a bird hide. A permanent canopy to keep rain and sun off is highly recommended, and can make the jetty into an attractive outdoor living area in summer.

Barriers may be needed to keep livestock or humans out of sensitive zones such as breeding and nesting areas. Humans are easily put-off by strategic barriers such as shallow water and very boggy soils, or more obvious obstacles such as sedges with sharp-edged leaves. Cattle will push through either of these, and will need to be kept out with conventional stock fencing.

Stock can sometimes be run in wetland situations without excessive damage, and may even have their uses as a management tool, but they will reduce plant diversity (photo 16), particularly in small areas. A fenced wetland can still be used for stock water by pumping or siphoning water to a stock trough or by including a fenced-off ramp lined with coarse gravel (photo 17) to control their access to the wetland. Cattle manure in moderate amounts is a useful source of nutrients for both plants and plankton, but they also erode the sides filling deeper areas with the resulting sediments. The constant presence of cattle may also encourage a build-up of liver flukes. There is some evidence that cattle drinking piped water from troughs may gain up to 20 per cent live weight over those drinking direct from a wetland or dam.

Livestock are not the only source of sediments in wetlands. Even average periods of run-off can cause problems if the soil upstream is exposed to erosion, and shallow wetlands can silt up very rapidly. Incoming water should first pass through some kind of silt trap, where its velocity is reduced enough that a large part of the silt will have a chance to settle out. Silt traps can be cleaned out more readily and less expensively than the wetland itself when this becomes necessary.

A small, relatively shallow settling dam above the wetland is adequate in many cases, but there are also many more elaborate designs depending on expected flow rates and how tightly incoming water can be channelled. Local soil conservation authorities and Landcare groups should be able to advise what is most suited to your area and conditions. Unfortunately, silt traps don't do much during flood periods, when they would be most useful, so even the best planned wetland will gradually silt up just as natural ones do.

At other times of the year, water may be in short supply and the wetland will dry out. If planted and designed along the lines of natural wetlands, this is no problem as it is a part of the annual cycle. There can even be advantages to cyclic rises and falls in the water level, including the control of some types of plant growth in the shallows. Wading birds will also have access to exposed, drying mud as water levels drop; this is an important foraging zone.

If you want to maintain the maximum water level at all times for appearance's sake, a reserve water supply somewhere upstream or within easy pumping distance will be needed. This should hold *at least* enough water to top up evaporation losses in the wetland in a fairly dry (not average) rainfall year, plus allow for evaporation losses from the reserve itself. However, the potential salinity concentration problems mentioned earlier apply to reserve waters as much as to the 'main' wetland itself.

Islands are usually carved into the wetland soils, but a raft in the form of a floating island may be used if the slopes are too steep to make this economical. I have yet to see a satisfactory, natural-looking design for these, and they don't seem to appeal much to birds, except as an elaborate and expensive resting spot. Floating islands need to be anchored against wind and current movement to keep them from running aground. If they sink, rougher ones can become a menace to swimmers and diving birds, although they will provide shelter for some underwater animals.

Floats used for such islands must be corrosion proof; even 44 gallon drums will last for decades if they are dried inside and out, painted liberally with a swimming pool paint, and then properly sealed. Wide diameter PVC pipes (sealed at all joins and ends) are expensive but easier to use as they need no pretreatment and can be more uniformly arranged beneath the island to float it without tilting. Less waterproof materials such as foam boxes won't hold an island up for long as they waterlog and lose buoyancy.

If plants are to be grown on a floating island, it should hold a soil layer about 10 centimetres deep, the lowest parts of this being underwater so that the soil remains wet at all times. Weigh a *wet* soil sample of known size to estimate how much extra weight it will add to the island; don't work from dry weight or your island may sink when wetted! Soil is best shovelled on and wetted once the island is already afloat. If this isn't possible, keep the soil dry until the island is launched, as wetting can double its weight. For islands with the top substantially above water, a gently sloped ramp may be needed for waterbird access to the top or it may not be used at all.

Islands made of earth are more attractive, but are often made so that they look like a miniature volcano protruding above the water. As most waterbirds aren't keen to struggle up steep slopes, islands should be flattened just above maximum water level, or can even be shaved to just below the water surface for birds that prefer to build raised nests in shallow waters. If the underwater slopes of the island are steep, a gentler landing ramp should be cut in along at least one side.

Such ramps can be in the form of a shallow bay cut into the centre, creating a horseshoe shape. This is particularly recommended for larger islands, as reports from the USA indicate that foxes may make their nest burrows in the drier soils at the centre of these. The extended curve of horseshoe-shaped islands also seems to separate potential breeding sites in the eyes of some birds, so more pairs may be willing to breed closer together as they may feel less crowded nesting on opposite arms of the bay within the island.

In extensive, shallow wetlands, piles of unwanted timber, stumps and rocks can be heaped to form inexpensive islands. Timber is not suitable for deeper waters as it will drift before sinking in random places. Other, relatively wholesome detritus such as spoiled hay can be added over the years to such islands. Its decay will boost populations of smaller invertebrates, attracting more birds and indirectly increasing populations of many other aquatic animals.

The underwater landscape should be planned in advance, including planting shelves and plateaus that will be submerged later; smaller planting pockets and narrow shelves can be done by hand even after flooding. If a good view over open water is desired from some places, the shoreline should drop quickly into relatively deep water here, planting shelves should be narrow, and lower-growing plants should be used. For extensive stands of reeds and rushes elsewhere, make underwater plateaus broad.

Shelves, plateaus and pockets can also be used as a management tool, for example to keep more invasive plants to one area only (assuming that these don't seed too widely). They should be at the appropriate depth for the plants selected, allowing for as much as 10 centimetres of prepared soil to go on top, plus a layer of sand or fine gravel if this will be needed to keep the soil in place. On very gentle slopes that aren't exposed to much wave action, constructed shelves aren't necessarily needed as there will be a range of depths already available.

• WIND, EROSION AND SURROUNDING VEGETATION

Many wetlands are wide open as well as flat, so wind can build up considerable velocity over them and generate substantial waves. These waves cause two types of problems: they limit the variety of plants that can be grown in some areas; and they can cause massive erosion (photo 11), particularly along retaining walls, and may ultimately destroy the wall itself. Wave action can be reduced by the strategic placement of islands, shallow areas and breakwaters.

Large, open expanses of water don't provide as much useful habitat or water treatment area as a series of smaller ponds or a large wetland broken up by underwater ridges near the surface (photo 12) so that plants can form wind breaks and wave breaks along them. Reeds, sedges and other taller water plants will help break the momentum of wind across the water's surface, but may increase water loss to the air as some species suck water up even faster than evaporation does. If the wetland is broken into many smaller, linked ones, taller windbreaks can be planted between and around the ponds, above the limits of wetland vegetation.

Belts of terrestrial vegetation around the wetland reduce sediment run-off from surrounding land, and help absorb excessive nutrients and contaminants that can cause water quality problems. A further benefit is reduction of noise and screening of movement that may otherwise disturb wetland animals, such as from nearby traffic or farm machinery. However, vegetation belts used as screens around a wetland may need wide gaps in them so that birds can see their way to fly in, especially if the water area is small.

Trees and larger shrubs shouldn't be planted along water retaining walls, partly because they may blow out in high winds, tearing part of the wall out with them. Their roots may also push through the retaining wall in search of water, creating a seepage line that will eventually breach the wall. Dead roots do even more damage as they may eventually rot away to leave a pipeline, and even smaller shrubs can be a problem for this reason. If you intend to plant along the retaining wall, choose species that prefer drier conditions so their roots aren't even tempted to stray into the wetter zones.

Orientation of a wetland to prevailing winds is important. In coastal areas of southern Australia, wetlands open to the south are most affected by south-westerlies, which can reach gale force at times. Towards the tropics, winds are more seasonal and blow from different directions during monsoons and during the dry season. High, open sites are the most difficult to deal with because they may be affected by wind turbulence as well as direction. Regardless of direction, strong winds generate waves. These waves can become higher and more destructive if they are allowed to build up over a greater distance, so the longer axis of any open water area should be across the prevailing wind direction, and not along it.

If it is too late to do anything about the site of a dam or wetland, a series of breakwaters may be used to absorb wave energy so that waves can't build up too much momentum. These are basically structures that redirect wave energy into the floor of the wetland. As most of this is concentrated in the top metre or so in inland waters, the solid part the waves hit doesn't need to be much deeper than this (but allow for evaporation losses in summer as well). Breakwaters must be firmly braced against the wave direction so that the energy of the waves hitting them is transmitted into the floor of the wetland through an angled brace behind. Floating baffles won't work, however securely they may be anchored, as they just move with the waves and barely absorb any of their energy.

Exposed planting situations will also need protection against wave action in the form of solid barriers of rock or concrete. However, there will still be considerable turbulence behind any such barrier if waves break over it, and a layer of heavy gravel over the planting soil will reduce washing out. The combination of waves and gravel limits the choice of plants in such situations to taller and more vigorous species that are already adapted to strong currents and disturbance.

If the wetland is still in the planning stage, it may be possible to weaken the impact of waves by having a shallow area (rather like a sandbar) a bit further out from the shore. This should be protected by rocks large enough that they won't be shifted by the expected wave energies. Breaking waves over such 'rockbars' can be quite dramatic in appearance, but much of their energy is lost here. The resulting reduction of impact energies at the water's edge means that finer gravels or coarse sand can be used on the surface of the planting beds, and allows more choice in planting.

One form of erosion rarely planned for is burrowing, the two most common culprits being freshwater crayfish and water rats. If natural wetlands nearby have either of these, they will inevitably turn up in a new wetland sooner or later. There is little point in trying to keep them out; after all, what is habitat if not for these and similar beasties? Just make sure that their energies are directed away from critical areas, particularly retaining walls where burrows could cause structural problems.

If you expect such immigrants, layers of shade cloth or other strong, fine meshes buried a little under the surface will stop them from burrowing, and will last indefinitely protected from sunlight in this way. If crayfish are likely to be the only problem, a gentle gradient no more than one in four on retaining walls will discourage digging, as they prefer steeper overhangs for their burrows. Crayfish are also less likely to burrow in wetlands where water levels don't vary much, although they will still cause some minor damage around the waterline.

• HABITAT

Many of the design considerations already looked at are directly related to the habitat needs of wetland animals, but the *idea* of habitat should be considered in its own right. All too often, a wetland is imagined to be a successful habitat regardless of what is in it or what was intended to be in it. This is particularly clear on sites where black ducks, *Anas superciliosa* (photo 13), are sometimes seen, but little else. These birds are often cited as evidence that the site is a successful habitat, however poorly it may actually have turned out.

If black ducks are treated seriously as an indicator of wetland quality, then even a muddy, unplanted farm dam would have to count as successful habitat. This bird is a widespread, migratory opportunist; it is even found in large numbers in municipal lakes, apparently thriving on a diet consisting mainly of white bread! When conditions are no longer even to their liking, black ducks can simply move on, an option that is not available to many other wetland animals.

A wetland habitat should be designed to meet the needs of specific target species, whether this is one animal or several compatible ones. Only then can it be known if it has been successful or if it has failed. Designers are often afraid to spell out exactly what they are aiming for, so it is worth pointing out that well documented wetlands that aren't as successful as planned can still add to our understanding of how they work, as so little concrete information of this kind is available. And, a well designed 'failure' may provide unexpected opportunities for other animals to move in, which will also add to our knowledge if everything has been properly documented.

Each different animal has a different set of requirements that must be met if a wetland is to provide complete and adequate habitat for it. Wetland animals are diverse, so it is obvious that no sweeping generalisations can be made about what is and what is not desirable in a design. The first step in planning for any particular species must be to learn as much as is possible about its life, habits and preferences from all sources available.

The common snake-neck tortoise, *Chelodina longicollis*, is as an example of a species that is apparently thriving in some wetlands yet may actually be at an ecological dead-end. This tortoise probably lives for many decades, and substantial populations are found in many inland waters of south-eastern Australia. A closer look often shows many mature animals but very few young ones. One survey of the breeding sites of freshwater tortoises has found that up to 94 per cent of nests in the study area were dug up and destroyed by foxes, a further 3 per cent being found by native predators such as goannas. How this compares to pre-European days is not known, but it certainly suggests that young tortoises are likely to be much less common than they used to be.

A priority for tortoise habitat must therefore be breeding areas secure from foxes. Females may also end up laying in water (where the eggs drown) if they are disturbed while trying to dig a nest, but fox-proof fencing (see Chapter 4) will also exclude humans and livestock. Their preferred nest sites are usually sunny places with reasonably soft or sandy soil a fair way up from water, and are often a surprisingly long way inland. A site that tortoises have seen flooded even once is unlikely to ever be used for nesting again.

There are obviously many ways in which these requirements can be met, with varying degrees of practicality. Fox culling may be most economical in large areas where fencing would be too expensive or difficult to patrol. This may not even need to be done every year, perhaps just during particularly good breeding seasons or every second or third year. Where disturbance by livestock or humans is also likely to be a problem, fencing is the better option, but it must be well placed to minimise costs while maximising protected area. For example, a study has shown that some North American freshwater tortoises will nest as far as 275 metres from water, but that a fenced buffer zone of 73 metres would take in 90 per cent of their preferred nest sites.

Fortunately, the other habitat requirements for snakeneck tortoises are modest. They will hibernate in or out of water, and will even feed happily on the introduced gambusia, a fish that has caused much havoc among native fishes and frogs. As long as their nests are adequately protected, it is likely that these tortoises will thrive in many created wetlands. Providing adequate habitat for many other animals may be just as simple if we have taken the trouble to familiarise ourselves with their life cycle and needs.

• MODIFYING EXISTING DAMS

If you already have dams on your property, they will usually have been made to hold water for stock or for irrigation, and you will probably need to keep on using them for the same purposes. There are many steps you can take to improve the value of such dams as habitat for a wider variety of plants and animals. Whatever changes you may decide to make, avoid tampering with the basic structure of a successful dam wall: it is all too easy to create leaks and weaknesses where there were none before.

If you keep livestock of any kind, one simple improvement is to fence them out, which will allow you to plant around and in the dam. Complete fencing is easiest to arrange, assuming that a siphon line can be put in to fill a drinking trough nearby. Even fencing stock out from a part of the dam will improve water quality, and if you close off access to the retaining wall this will stand much longer. However, it is possible to manage stock in wetlands with satisfactory results under some conditions, particularly if you are concerned about maintaining a diversity of invertebrates in more ephemeral wetlands. Many of these smaller creatures will thrive where livestock keep vegetation down, and will benefit from the resulting fertilisation by manure (photo 19).

Existing islands and peninsulas that are too steeply sloped for waterbirds should be levelled to just above the highest water level. This can be done by a bulldozer if a shallow and not-too-boggy access area is available, or an excavator if the island is not too far from shore. Hand tools do the neatest job, and should be used to trim up after the use of machinery; if hand tools can be used for the whole job, it will be easier to leave any existing patches of vegetation unharmed.

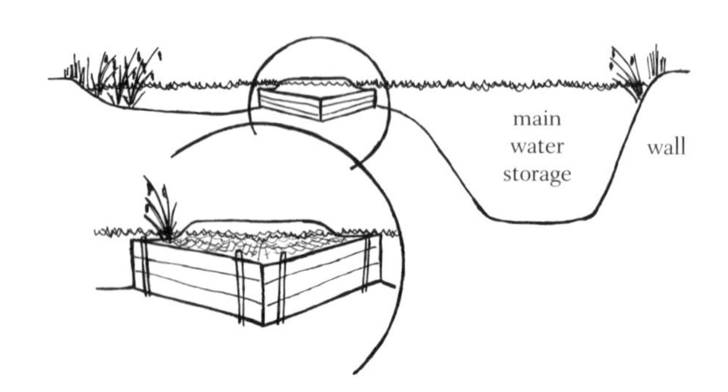

Cross-section of a dam showing planting plateau behind and a flattened island terraced under water.
Inset: terracing filled with soil from the levelling of the island.

Steep-sided islands and peninsulas are usually also steep underwater, and submerged terraces around them are a useful way of increasing planting area. The extended shallows created this way also make access to the shore easier for waterbirds and reptiles. Terrace walls that stop soil sliding off into deeper water can be made from old lumber, which can be very long-lasting underwater, or with larger rocks, waterlogged tree-trunks or even slabs of concrete rubble. If the timber is held in place by star posts, the tops of these should be hammered until rounded as they are otherwise dangerously sharp.

Planting shelves can also be put into an existing dam by a backhoe or an excavator even when it is full, but must never be dug into water-retaining walls. Perhaps the most dramatic improvement you can make to an existing dam is to add an extensive shallow planting area adjacent to it to form a new wetland in its own right. Properly planned, this will encourage a wide range of new arrivals, from nesting birds to smaller underwater animals.

Such areas should slope gently down towards the dam itself, although they can also be done as a sort of artificial billabong adjoining the dam and only connected to it during times of highest water levels. A shallow, graded area like this will allow a wider variety of plants to grow than the steeper sides of the dam itself. The beauty of a properly designed combination of new wetland and old dam is that you can use the dam much as before, including for irrigation, which is usually only needed once most water animals and birds using the wetland have already completed their natural breeding cycles.

The dam will fill with seasonal rains, overflowing into the wetland and triggering plant growth and animal breeding behaviours just as in comparable natural wetlands in the area. Once run-off stops, evaporation begins, slowly at first but accelerating as the sun develops a bit more bite. By carefully choosing the depth of the new wetland sloping to the dam, you can decide approximately when the new area will become completely dry, so that the dam will have retreated back into its original area again. For example, if evaporation in shallow wetlands in your area will draw off about 15 cm of water within 9 weeks of rains finishing, and you need to start irrigating at about that time, then the deepest part of the new wetland extension should be about 15 cm deep.

Using the same example, if you plan to start irrigating at 6 weeks, the extension should be about 10 cm deep at the most; if you won't need to irrigate until 12 weeks, it can be up to 20 cm deep. These depths should be measured from the top of the planting soil, not from the bare clay it rests upon. Although the extra surface area added by the wetland will increase the *total* evaporation loss, any extra loss will come entirely from the extension. Remember that the dam would usually have lost pretty much the same amount of water by evaporation in the same time, anyway, even without an additional habitat area built on behind.

Whatever depth you decide is most compatible with your other planned uses for the dam, use natural wetlands of the same depth in your area as a guide to planting. A wetland dominated by *Juncus* species, which usually grow where they are high and dry within a month, won't attract nesting birds that need water around their nests for 2–3 months. However, it will attract those that need to forage over drying mud, and even if their nests are somewhere else in some other wetland you will have increased the total amount of habitat available to them. On the other hand, a wetland area dominated by plants that grow mainly where water will stand for many months gives different signals to potential breeders that recognise suitable breeding habitat at a glance.

• WATER TREATMENT WETLANDS

Wetlands are not a miracle cure for wastes running off surrounding lands, but they certainly help to improve the quality of water moving through them. This is partly done through the action of microorganisms and partly through luxury uptake of nutrients by plants beyond their actual needs at the time. A complete discussion of water treatment is far beyond the scope of this book, but a brief outline of design considerations for this purpose would not be amiss. It is worth noting that many different water treatment designs have been used worldwide, but none of these have proven to be so far ahead of the others that they are likely to set any standard. Most designs are variations on just a few basic themes, and it remains to be seen which will continue to be created and used in the next century.

Some good research towards workable systems has been done in the past few years (much of this overseas), yet many designs being produced show little sign that their perpetrators have kept up with recent developments. Larger projects are often put together by the committee approach, where several consulting organisations are asked to collaborate. Each of these organisations will usually have a well earned reputation in other fields, although that doesn't mean that they will be able to produce a coherent wetland design. An engineering firm will be called on to produce the overall design, a landscaper will be in charge of aesthetics, a botanist will be asked to produce a list of plants known to occur in the area, and a nursery will be required to propagate and plant an apparently randomly chosen assortment of these plants.

There are weaknesses with such a system and the results produced can be unfortunate. I have even seen a wetland that needed irrigation systems to keep the inappropriately chosen plants alive at times! In another case, a channel planned and planted to improve water quality downstream created an erosion problem and increased sediment loads instead. Such examples aren't all that rare, and this problem is not confined to Australia alone. It is not all that long since a well funded program in the UK created wetlands at a 4° angle to increase the flow-through rate. The water certainly flowed through quickly, but not enough remained to keep wetland plants alive.

Many water treatment wetlands are also too small for the catchment area or runoff volume they are expected to service. A 1 hectare wetland will have little effect on water running off 100 hectares. Unless the wetland area is large compared with the catchment area being serviced, it may be necessary to recycle water by pumping to increase the time available for treatment. Assessment of the effectiveness of water treatment is often neglected, which may make it impossible to determine whether the wetland has had any real effect on water quality.

The most basic type of treatment wetland is filled by surface flows, where water runs through a shallow basin that is heavily planted. The flow should be as uniform as possible through all areas, and various combinations of baffles and channels may be used to keep it that way. Such wetlands must be precisely levelled, and in large-scale work laser grading is used for accuracy. On a small scale, flood the area to be used shallowly once it has already been roughly levelled. The water surface can be used as a benchmark for levelling with a hand-held hoe; this is probably how even large flat areas such as the bases of the Egyptian Pyramids were done. On a sloping site, a series of smaller, narrow channels flowing downhill from one into the next can act like a larger wetland, and the channels will reduce the need for baffles to maintain uniform flow.

Some wetlands use an open substrate for plant growth, with water flowing beneath the surface only. This is particularly useful in areas where mosquitoes are a problem as it leaves no open water in which they can breed. The ultimate extension of subsurface flow wetlands is to simply use the water to irrigate terrestrial plants, particularly tree plantations. This is a favoured method for disposing of water which has such serious quality problems that it should not be allowed to return into natural waters.

Some promising systems allow water to flow downwards through a loose substrate such as sand densely packed with living roots of various reeds and rushes. These vertical flow wetlands have mostly been used for small-scale water treatment as it is difficult to keep vertical water movement uniform over larger areas. As the root zone retains reasonable oxygen levels with this type of flow, wastes are broken down more rapidly than in oxygen-poor environments.

None of these designs will cure serious water quality problems such as heavy detritus loads, industrial contaminants or excessive manure and organic waste levels, so pretreatment to filter or sediment out unwanted materials out may be necessary. No matter how good the pretreatment, sediments and nutrients will build up gradually in every wetland as well as organic material from the plants themselves. All water treatment wetlands therefore have a limited working life, and are likely to need at least partial scouring and replanting every decade or two, depending on how quickly silt and decay materials build up.

Finally, it should be noted that effective water treatment wetlands offer a fairly restricted kind of habitat. They are shallow, show little variation in flow rates, and are densely packed with a (usually) limited range of plants. Certainly, animals specialised for life in dense reed beds — such as bitterns, rails and many insect larvae — will thrive here, but species diversity is rarely as high as in more varied wetland systems.

3 CONSTRUCTION AND LAYOUT

Laying out and constructing wetlands is not particularly difficult, assuming that you have access to the right types of machinery, skilled and sympathetic operators, and suitable soils. It is usually necessary to be present for the more critical stages or the work may not end up as planned. Many operators are resistant to methods that are novel to them, and will take every shortcut possible even when being paid by the hour because they only believe in the methods they have always used. This is why one in three Australian farm dams leaks; the figure would be higher except that some soils will hold water no matter what you do to them!

Bulldozers are most commonly used in the construction of dams and wetlands, but other machines may be more appropriate for many sites (photo 24). Your initial surveys should aim to find places where a minimum of earthmoving will create maximal area of wetland; for example, by blocking narrow gullies with extensive, undulating areas behind. However, suitable water-retaining soils are also important, and should ideally be under the intended site or reasonably close to it.

A ground survey with just a surveyor's level may be all you need to set out future wetland boundaries, and most earthmoving contractors will have one of these available. However, for very large areas it is sometimes useful to be able to work initially from an aerial photo. These may be available from government authorities, but private pilots sometimes photograph a whole district and then advertise in the areas they have covered. Technology is now available to lay in contours as small-scale as 1 metre over such photos, although the process is still fairly costly. Contoured photos may help you avoid making mistakes such as drowning trees and shrubs that turn out to be unexpectedly near areas to be submerged.

• SOILS AND CONSTRUCTION

Common sense will solve most problems that come up during excavation into a water table, but construction of a water-retaining wall requires the specialised methods of dam construction. The first step is to check whether the clays available are suitable or will need special treatment. Moist clay that can be rolled to the thickness of a pencil without breaking apart will hold water once it has been compacted. Dispersive clays tend to break up because their particles don't stick together properly when wet.

There are varying degrees of bonding in dispersive clays: weakly dispersive clays may make a satisfactory wall without much special treatment, but it is best to obtain specialist advice for strongly dispersive clays, which dissolve readily in water to form electrically charged particles that can only be settled by changing the electrical conductivity of the water. This is usually done by adding electrolytes such as lime (particularly in the form of gypsum) or by otherwise increasing the hardness or salinity of the water.

Waters containing highly dispersive clays will clear when the total hardness plus salinity is approximately 250–300 parts per million. Curing a muddy water problem will also indirectly fix the water-holding ability of the wall; this is why clear dams don't leak appreciably if they have also been properly compacted. Only disturbed dispersive clay soils are likely to seep or tunnel through walls. The same material left undisturbed below the original soil level generally holds water perfectly well. Untreated dispersive clays can be used to make up the bulk of a water-retaining wall as long as there is a layer of good quality water-retaining material either lining the inside of the dam or as a core inside the wall.

Once the nature of the clays available is known and allowed for, the next step is to work out how much material is available, and where. A rough estimate can be made with a hand-held 10 cm diameter earth auger, sampling in a few places over the planned site. These are only about 90 cm long, but can be extended another couple of metres with additional sections of steel pipe if you need to auger deeper for suitable material.

The first step in actual construction is removing all topsoil from the site, and stockpiling it nearby until the water-retaining structure of the wetland is finished. Layers of the topsoil can then be spread back over the wetland, mixing in any additional materials (fertilisers, organic matter) at the same time, as required. Even a bulldozer can do this quite efficiently. It is often recommended that the topsoil be replaced uniformly over the whole area of a wetland or dam, but it is wasted in deeper waters as few water plants will grow much below a metre except in the clearest waters. Unless you have plenty of topsoil to spare, save it for the shallows.

• CONSTRUCTION WITH ABUNDANT MATERIAL

The whole retaining wall can be made of suitable water-retaining material if there is no shortage of this. Before the wall is built up, the subsoil from which it will be started must be scarred and loosened slightly to a depth of around 5 cm, allowing the base and wall to bond together more firmly. (If the wall is simply built up from a smooth surface, water may seep through between the two faces and the whole wall may slip in extreme cases.) The tines on the back of a bulldozer are commonly used for this scarring, after which wall materials can be spread into place layer by layer.

There is no such thing as a *completely* water-tight earth wall. Even well compacted, high quality clays seep, however slowly. The greater the pressure they are under, the greater this seepage will be, and the more care must be taken to keep this to an acceptable level. Pressure is directly related to water depth, not to the total amount of water held in a dam or wetland. At its base, a retaining wall with water 10 cm deep behind it holds one-tenth the pressure compared with the base of a wall with 1 metre of water pressing against it. A wall holding back 1 metre depth of water has the same pressure on it whether the wetland is 100 square metres or 100 000 square metres in area (photo 20). However, if the area is large, the wall will probably need protection from wave erosion even though there is not much actual pressure behind it.

Once the wall is fully built up or close to it, the inside and any other uncompacted areas that have been stirred up during building must be rolled to reduce seepage to a minimum. For walls holding back 3 metres depth of water or less, a single rolled layer (six or more passes of a roller) should be enough. For deeper waters, each further 3 metres of depth requires a further loose layer of material approximately 30 cm deep. This should be spread over the previously rolled surface, and be rolled again separately.

Thus, in a dam 9 metres deep there should be three separately rolled layers in the deepest parts. Two layers would be adequate in those areas that are less than 6 metres deep, and one layer in areas only up to 3 metres deep.

• CONSTRUCTION WITH LIMITED MATERIAL

If there isn't enough suitable material to make up the whole retaining wall, what there is should be saved to make a skin or a core, and the bulk of the wall must be built up of whatever other soil is available. If there is a good layer of water-retaining material underlying the whole area of the dam or wetland, a core of the same material running down through the centre of the wall will keep water in. This must be dovetailed into the underlying layer by putting in a trench at least 30 cm deep from which the core will be built up. The taller the wall, the deeper and wider this trench should be.

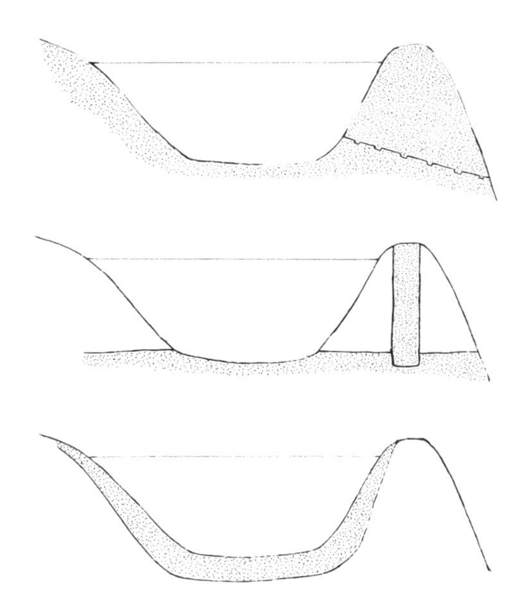

Three types of water-retaining walls, which are each used for different conditions. At top: a wall made entirely of good quality, water-retaining soil (shaded) where this is abundant. The subsoil has been ripped to 5 cm depth to allow closer bonding with the wall material. Centre and below: alternative constructions with limited good quality soils, the bulk of the wall being made up of more porous soils (not shaded).

The core is just a column of water-retaining material built upwards from the trench, with sloped supporting walls on either side made up from whatever other soils are available; these are built up at the same time as the core. Unless the core is made from high quality material that doesn't need compacting to give a water-tight seal, it should be built up in layers of about 30–40 cm depth, each of which is rolled before the next layer is spread on top.

If there is no suitable layer of water-retaining material under the site, the basic shape of the entire wetland or dam can be carved out of whatever soils are available. It can then be lined with a skin (or blanket) of water-retaining material. This skin should be thickest in the deepest areas, up to 60 cm in water 3 metres deep. It can taper to as little as 30 cm in the shallows where there is little pressure, if there is no problem with tunnelling by freshwater crayfish. The skin should be rolled, with an additional rolled layer 30–40 cm deep being laid for every additional 3 metres of depth.

• ALTERNATIVE MATERIALS

Not all sites have good water-retaining material available. Some soils can be modified to hold water, for example by the addition of lime, sodium tripolysulphate or other chemicals that will improve their structure. Most such alterations require expert advice on the type, dosage and application of each chemical, although you can guess how much lime to add to dispersive clays if you don't mind the extra cost if you overdo the amount. Random amounts of other additives won't do any good, and may even prevent the soil from ever holding water.

Water-retaining materials can also be brought in from elsewhere to line a dam or wetland. Usually, the simplest and least expensive of these is better quality clay applied to form a skin, as already described. There are few properties where suitable clays can't be found somewhere nearby. Various types of synthetic liners are available for dams, and these vary considerably in cost, strength and expected lifespan. Tough, long-lasting liners such as EPDM membranes can be joined to cover any area but are very expensive; butylene liners are comparable in quality but cost around twice as much. Liners must be covered with soil to give a natural bottom. This is generally done by machine, and cheaper liners will be fairly easily damaged during this stage.

Bentonite clays are another expensive solution, although they can be applied quite thinly in shallower areas. This is usually bought as a dry powder that expands dramatically when wetted, increasing in volume 10-fold for calcium bentonite and as much as 17-fold for some types of sodium bentonite. Bentonite is laid as a dry blanket under a layer of top-soil, swelling when wetted to form a layer impervious to water. Straw, leaves and manures are also sometimes used to try to seal sandy soils with a layer of well rotted organic matter called gley, an effect similar to the peaty bottoms of lakes that form among sand dunes (photo 18). However, this method is rarely used in Australia except where such materials are available locally in considerable amounts and are inexpensive.

• MACHINERY

Deciding on the appropriate machinery to use for earthmoving is usually simple once the wetland has been laid out and the materials available are known. Combinations ofmachines will be most effective in many cases. For example, a small bulldozer and an excavator have maximum manoeuvrability for shaping a small wetland. A large bulldozer and a scraper are more appropriate for creating a large wetland, the bulldozer speeding the scraper's work by loosening soil and pushing this into its path.

EXCAVATORS AND BACKHOES

These are digging machines with a long arm that finishes in a bucket for lifting soil. Backhoes are smaller machines, and are only useful in very small areas because both their reach and bucket size is limited. However, they can do some reasonably fine detail work, and shift rocks and logs around during landscaping (really precise trimming can only be done with hand tools, although this isn't practical over large areas).

Excavators (photo 21) are more powerful, with a much longer reach and a much larger bucket. They are also mounted on a caterpillar tread that allows them to work in places where a backhoe can't go. They are particularly useful for digging below a water table and for contouring shorelines. They can also be used for detailed work in wetlands constructed by bulldozers or scrapers, neither of which can create fine channels, or for making smaller ledges and planting pockets. Don't excavate directly into wet areas where water flows for even a part of the year, as this can cause serious erosion problems. Deeper pools around such areas are best dug well to the side of the main flow, where they will be filled by overflow.

BULLDOZERS

These are most useful in general earthmoving (photo 22), and come in a wide range of sizes suited to many different jobs. The smaller dozers are good for landscaping in skilled hands, while larger ones are more efficient at moving earth in terms of cost per volume shifted. The caterpillar tread they run on is designed to spread their weight, not for compaction, so they should be used in conjunction with a sheepsfoot roller or the walls they build up will often leak.

SCRAPERS

Basically a specialised truck (photo 23), the scraper's 'belly' is a gouge that can be lowered to scrape up soil or clay so that it can be shifted to another area where the contents will be spread. Scrapers work most efficiently over larger areas where they can follow a continuous loop, picking up soil in one area and dropping it in another, and circling back to pick up another load without slowing down unnecessarily. They are fairly heavy themselves, and as they carry anything from 5–10 tonnes of soil as well (all this concentrated in the small contact area of tyre and ground) they give very good compaction from passing over the same areas repeatedly.

ROLLERS

On sites where only a bulldozer is used, compaction is usually poor unless a roller is also used. Soils with as little as 20 per cent of a suitable water-retaining clay can hold water if they are properly compacted while they are reasonably moist using a sheepsfoot roller. These are heavy cylinders running on many spiky feet, and are passed over the same area six or more times for best results. In the early passes, their feet sink in to the hilt. By the last pass, the roller should be tip-toeing around on the ends of its feet, and the ground may be so hard that a shovel will ring if struck on the finished surface.

Over smaller areas, a detachable roller can be pulled by a four-wheel drive tractor. Although a bulldozer can easily pull a smaller roller like this, it is uneconomical to couple and uncouple frequently, and bulldozer time is more expensive than tractor time. Over larger areas, a self-propelled sheepsfoot roller (photo 25) should be used. These usually also have a blade for fine trimming, and vibrate so that compaction is more efficiently done.

LASER GRADING

On sites where precise control over flow rates and directions is necessary, especially in larger water treatment wetlands, lasers are used to determine exact gradients. This technology is already widely in use on irrigation farms and in rice growing areas. Specialists in this type of work are usually highly skilled, and don't need to know much more than boundaries and gradients to set out this type of wetland.

• WATER CONTROL

All wetlands and dams that have water running through them at any stage must include ways of controlling water going in or out of the system. The most basic of these is an adequate overflow (or bywash) that allows water to escape during wetter periods to ensure that the retaining walls are not breached by water pouring over them. Where the total catchment area draining through a wetland is small, an overflow can be as simple as a concrete pipe or two passing through the wall. Where the catchment is large, the overflow must be broad enough to carry flood waters so that these don't bank up too much and go over the wall itself.

Overflows are usually situated to the side of a retaining wall, so that the escaping waters can be directed away from the wall itself to prevent undercutting. The overflowing waters should be directed along a ramp sloped as gently as possible and held together by dense plantings of appropriate plants, heavy gravel or even concrete if the slope must be steep. Overflows that can be opened and closed when necessary can help reduce flooding downstream in periods of heavy rain by retaining some of the water for subsequent release when the main flood has already passed.

Even at drier times the ability to control the drainage of water from a wetland can be useful, especially to empty it if required. This is called drawdown. Although little used in Australia at present, drawdown is likely to become an increasingly useful management tool, especially in the control of weeds and vermin. There are two parts to an effective drawdown system in constructed wetlands and dams: a sloping floor or sloping drains leading to an internal sump (photo 26), and a way of removing the water when necessary.

Sloping floors limit the ways in which a relatively shallow wetland can be designed because they create a regimented arrangement of slopes that may not fit well with other plans. A simpler variation is drainage lines across the floor of a wetland. These should slope slightly downwards as they run into deeper channels, all of these leading ultimately to a sump. If the wetland is drained, this sump will be the last remaining pool of any size on the floor, and will be easy to net when rescuing native fishes or to lime heavily to kill unwanted exotic species.

Water can be pumped out, siphoned out, or allowed to drain along a permanently fixed pipe passing through the retaining wall if there is one. Siphon lines may be relatively slow to drain a large volume, but have two advantages: they are less expensive than installing permanent pipes, and they don't potentially weaken retaining walls. However, they can only be used where the bottom of the wetland is not too far from an even lower point to which water can be siphoned off. Where a series of smaller wetlands is planned, one siphon may be all that is needed as it can be shifted around as required.

A simple siphon can be made from a piece of agricultural polypipe. The larger the diameter this is, the faster the flow rate will be as friction becomes proportionately less for the volume of water moving through. The siphon is weighed at the siphon (sucking) end, and has a valve to control flow at the other end. It can be filled with a pump, or can be pushed underwater section-by-section by hand to fill it gradually. The valve is shut once the whole line is submerged. The valve end is then pulled out and over the retaining wall to a level below the lowest part of the wetland before the valve is opened again to begin the draining. The weighted siphon end should be in the sump when the siphon is started.

Pipes can be built into retaining walls, but considerable care must be taken as they are a potentially weak point. The pipe itself must have watertight baffles welded, glued, or otherwise firmly clamped at regular intervals along its length. Water will tend to flow along the surface of smooth pipes even within the wall, eroding a tunnel, but there will be no flow if baffles are in the way. This type of tunnelling effect is a common cause of dam failure. Water will even move like this along a piece of fencing wire if it runs right through the wall.

Pipes through walls are usually set in a trench that runs below the base of the retaining wall. As this area will have more water pressure on it than any other part of the wall, the soil around the pipe must be compacted thoroughly. No machine can do this job without damaging the pipe; it must be done by hand, and a dirty, tiring job it is too! This is why I recommend using siphons instead wherever possible. Even a 35 mm diameter siphon can shift up to 100,000 litres per day given a reasonable fall, so only really large dams or wetlands need anything more elaborate.

• THE UNDERWATER LANDSCAPE

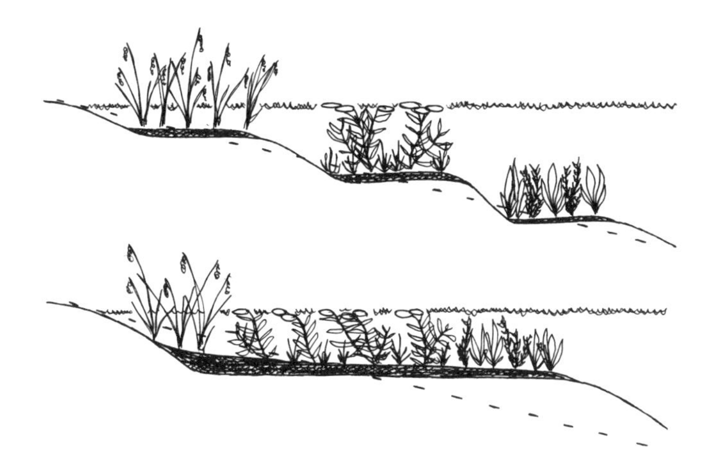

Planting shelves (above) and a plateau (below), showing two ways in which the same original slope (dotted) can be modified for different types of planting.

The arrangement and type of soils below the surface decides where, how and which plants will grow, in turn deciding the very nature of a wetland as these are a dominant part of many habitats. Suitable soils and growing conditions for plants are discussed in Chapter 4; however, the base over which such soils are to be spread is a part of the framework of the dam or wetland.

In a natural wetland, underwater slopes will often creep gradually out into deeper waters, but it is not necessarily a good idea to try to imitate this in a newly constructed site. Loose soils spread over a sloping base will tend to slip or wash into deeper parts of the dam or wetland, especially as it is generally uneconomical to plant them heavily enough to bind together quickly. To counteract this tendency to slippage, constructed planting areas should be level or even slope slightly away from the deeper areas. With time, the shoreward side of each level area will tend to build up with new material washed in or eroded by wave action, giving a more natural profile.

Planting areas can be loosely divided into shelves, plateaus and pockets. Shelves are often much the same width over their whole length (although they don't have to be), while plateaus can be almost any shape. Pockets are smaller, often dug in by hand, and are usually used as a way to separate plants or contain patches of more spreading species. The depth of any shelf, plateau or pocket should be chosen to suit the plants you wish to grow in that area, so you should have a reasonably clear idea of the species you intend to plant before construction has even begun.

There is no reason to keep shelves, plateaus and pockets always separate and clearly defined; a wetland is not a formal flower garden. However, for a more natural appearance the underwater landscape should be arranged to give some impression of zoning. For example, a shelf running along an edge and planted with no more than two or three species can be made to look much like the edges of many billabongs or other natural wetlands. A deeper shelf a bit further out with deeper growing species will help to extend the illusion. By contrast, plateaus of a fairly uniform depth will allow extensive stands of various sedges to develop, rather than narrow zones.

When planning planting areas like these, allow for the addition of a layer of soil. The final depth of a planting area isn't measured from the underlying clay, but from the surface of any soils spread over this. For example, a shelf base 15 cm deep with 5 cm of soil over it will give a planting area 10 cm deep. Deeper blankets of soil are better if enough is available as they will hold moisture longer than the compacted clays that often underlie them, but there is rarely any need for soil depths greater than about 15 cm. Shallower soils will dry out sooner once water levels have fallen below their shelf or plateau, so these are best planted with more drought-tolerant species.

4 PLANT SELECTION, PROPAGATION AND PLANTING

In general, wetland and aquatic plants have been grown and treated in much the same ways as terrestrial plants used in revegetation. However, their biology and ecology are sufficiently dissimilar that a rather different approach to many aspects of their selection, propagation and planting is not only warranted, but is also often more economical. These aspects are not just of interest to those who wish to propagate these plants but are also interesting in terms of what they tell us about the natural history of the plants, particularly under conditions at the present time.

• PROVENANCE

Perhaps the most basic rule-of-thumb used for selection of propagating materials in revegetation is to propagate from plants or seed obtained from as close to the planting site as possible. Such sources are referred to as being of 'local provenance', and are assumed to be well adapted to their area, certainly better than other sources of the same species from further away. As with many simple, commonsense concepts, the logic behind the idea comes unravelled in parts when looked at more closely.

The universal application of one generalised rule like this is perhaps a little too sweeping, especially considering that there has been virtually no research done anywhere in the world to verify it. It also ignores significant problems, includ-ing changes in climate and growing conditions since European settlement, species naturally in decline, and the increasingly limited genetic base of reduced and fragmented populations. Numerous volumes have been written on the genetic problems in particular, and the interested reader can find many months of fascinating reading on this and related areas in the references suggested at the end of this book.

There is also the problem of just what 'local' means, particularly as it has been interpreted rather stringently by some people. Some propagators will only use materials collected from within about 10 km of the planting site, while others invent their own guidelines. Some collectors even restrict their collection of planting materials to current council and shire boundaries, a bizarre and unbiological notion if ever there was one! The greatest danger with restricting the collecting area so much is that there may be very few populations present, and these will often be closely related and genetically impoverished.

2

4

3
5

9
10

12

13

14

15

16

17

18
19
20

22

23

24

25

26

27
28
29
30

31

32

33

34

35

39

40

41

42

43

44
45

46

47

48

49

1 A clear coastal waterfall supporting a diverse assortment of algaes, bacteria and mosses, and some unusual insect larvae. Rainbow Falls, Cape Otway, Victoria.

2 Flooded pool in peaty damplands of south-west Western Australia, surrounded by male and female tussocks of *Leptocarpus scariosus*. These would normally be well above water level, but are adapted to periodic flooding. Five of the native fishes of the area are recorded from this pool, as well as the crayfish *Cherax preissii* and the tortoise *Chelodina oblonga*.

3 An unpromising looking habitat, this roadside ditch is home to more than a dozen short-lived invertebrate species, including two relatively large crustaceans: *Branchinella australiensis* and *Lepidurus apus*. The water stands half a metre deep here, yet the ditch is completely dry for at least six months of each year. The grass is *Glyceria australis*.

4 A spring-fed swamp dominated by *Melaleuca squarrosa* with an understory of ferns, including *Todea barbara*, *Gleichenia dicarpa* and *Histiopteris incisa*. The canopy is so dense that ringtail possums were still active under it at midday in midsummer. A semi-terrestrial crayfish (*Engaeus fultoni*) is abundant in the waterlogged soils below.

5 Meanders and billabongs along the floodplain of a small river.

6 Active stream erosion, caused partly by the collapse of several willows.

7 Lake Ayr, Tasmania, with the shoreline dominated by tall spikerush (*Eleocharis sphacelata*).

8 Seedling grey mangroves (*Avicennia marina*) at low tide, with a seagrass (*Posidonia australis*) at their feet. Gudgeons, gobies and shrimps are common in the seagrass, while the crab *Sesarma erythrodactyla* is the most conspicuous animal among the mangroves.

9 This endangered gudgeon (*Mogurnda adspersa*) thrives in densely vegetated wetlands, particularly those free of the introduced fish Gambusia (*Gambusia holbrooki*).

10 A jetty used for swimming and fishing, with a canopy against sun and rain.

11 Even along the shores of natural lakes, wave action eats away at the shoreline. *Juncus kraussii* is seen here growing at the feet of swamp she-oak (*Casuarina glauca*), which will drown as permanent waters advance further through its root zone. Myall Lakes, NSW.

12 A view across one of many pools at Piccanninie Ponds, South Australia, separated by ridges covered in vegetation. The ridges prevent large waves from building up even though this wetland extends over many hectares. The varied underwater landscape includes limestone caves connected to the sea. Marine fishes can be seen here at times as well as several freshwater species and other animals, such as the tortoise *Chelodina longicollis*.

13 Black ducks use a wide range of foods and will even visit the most degraded farm dams. Their presence is not an indication of high quality habitat. The plant in the foreground is *Potamogeton sulcatus*.

14 Roots of a river redgum (*Eucalyptus camaldulensis*) left dry as waters fall in a billabong. The tree is well adapted to cyclic flooding, but if it is forced to stand in water for most of the year it will gradually die.

15 Walking ramp through a mature stand of lignum (*Muehlenbeckia florulenta*), allowing easy access except during extreme flood conditions.

16 A partly fenced farm dam shows the difference cattle can make to a dam or wetland. This amount of damage has been done by just six half-grown dairy cows.

17 Cattle ramp for watering access, lined with rock rubble to minimise erosion.

18 A lagoon among coastal sand dunes, which retains water because a layer of peaty organic matter (similar to gley) has built up in the sand.

19 This pond was originally grazed by cattle, which kept vegetation cropped, and supported diverse invertebrate populations including several unusual crustaceans. Since grazing was stopped 10 years before this photo was taken, *Typha orientalis* and *Cotula coronipifolia* have almost filled the previously open waters, the southern brown treefrog (*Hyla ewingi*) has moved in, and the previous invertebrate tenants have largely disappeared. Resumption of grazing at this stage would probably see the previous balance restored.

20 A handthrown diversion wall about 20 cm high used to redirect flow to another channel, and holding back an additional 200 000 litres of water on top of this dam's normal capacity.

21 Excavator.

22 Bulldozer showing tines behind.

23 Scraper.

24 Bulldozer and sheepsfoot roller tread marks, showing a great difference in penetration depth even though this clay is already fully compacted.

25 Self-propelled sheepsfoot roller.

26 The opening to a drainpipe built through the retaining wall of a wetland.

27 Sacred ibis nest in a clump of *Eleocharis sphacelata*. Many species of bird feed, breed or hide among such plants, and are likely to have seeds catch in their feathers.

28 Waterbuttons (*Cotula coronipifolia*) studded with yellow flowers, forming a dense mat around a low-lying stock waterhole.

29 *Isolepis nodosa* is sometimes found in wetlands but usually grows around their fringes or on higher ground. Its temperate southern hemisphere range is similar to that of waterbuttons, yet no one doubts that it is a native plant.

30 A water ribbon species (*Triglochin procerum* Western variant) growing profusely in the peaty, nutrient-poor soil of a coastal wetland.

31 *Lepironia articulata* forming a fringe around a lagoon, with a paperbark *Melaleuca quinquenervia* behind. This tall sedge produces very small amounts of seed for its size, ripening it over long periods of time so that collecting can take many hours.

32 *Baumea articulata*, one of many sedges that show varying and unpredictable degrees of seed viability between populations.

33 A saline drainage ditch that has been colonised by *Selliera radicans*, *Triglochin striatum* (terete form), *Samolus repens* and the grass *Distichlis distichophylla*.

34 *Blechnum minus*, a wetland fern readily propagated from spores.

35 Inexpensive propagating units for wetland plants: foam trays in a shallow pond, mostly using the waterlogged bog method of germinating seed. The taller box at right holds plants that are less tolerant to waterlogging.

36 Dramatic seasonal changes in water levels are normal in many types of wetlands, but the
37 greater these are, the fewer types of plants will tolerate such conditions. **(36)** An inland billabong seen full in late spring. **(37)** The same billabong almost dry in late autumn. In the foreground is a sedge (*Carex tereticaulis*). In the water, *Triglochin multifructum* and *Myriophyllum papillosum* disappear temporarily as levels fall.

38 Sacred lotus (*Nelumbo nucifera*), a spectacular, large-flowered tropical plant.

39 The floating fern *Azolla filiculoides* showing colour variations in a single clone.

40 A salt marsh showing striking zonation of plants with varying degrees of tolerance to salt and wet soils. In the foreground is red *Sarcocornia quinquenervia*, then the grass *Sporobolus virginicus*, the greyish rush *Juncus kraussii* where the ground falls away again, and a distant fringe of mangroves at the edge of the sea.

41 Paperbarks are common through much of Australia, growing where their roots may be flooded for weeks or even months but they do not tolerate permanent inundation. This is *Melaleuca rhaphiophylla* from south-west Western Australia.

42 *Monochoria australasica* flowering near the edges of a tropical lagoon, with the native waterlily *Nymphaea immutabilis* behind.

43 A settling dam overgrown mainly by the introduced *Typha latifolia*. The non-flowering swathe across the middle was sprayed to control it, but has only been set back enough to allow the plants behind to flower first. In the foreground, the native common spikerush (*Eleocharis acuta*) has not been affected at all. Chemical control of this kind is often ineffective in wetlands, yet has successfully killed all frogs and fishes in this dam. In a disturbed, artificial site such as this it is better to put up with any weeds that can't be controlled by hand if they provide adequate habitat for smaller animals.

44 The inland form of tall flat-sedge (*Cyperus exaltatus*), a common species in large areas of eastern Australia.

45 A coral fern (*Gleichenia dicarpa*) growing with its foliage in full sun, and roots around a spring-fed waterfall.

46 Two *Persicaria* species flowering along a stagnant creek. In the left foreground are upright plants of *P. praetermissa*. At right and on the far bank is the sprawling *P. decipiens*.

47 *Pycnosorus globosus*, seen flowering on a shallow floodplain after water has receded. An orb-weaving spider (*Eriophora sp.*) is abundant here, and mimics the older flowerheads among which it makes its webs.

48 Duckweeds include the smallest flowering plants in the world. These are full-sized plants of *Wolffia australiana*.

49 Cumbungi (*Typha orientalis*) well-entrenched in the landslide soils that dammed this natural lake.

Such restrictions may even make it necessary to propagate from a single stand if no others occur within the prescribed area. In the case of many wetland plants such as sedges, isolated stands may not just be closely related, they may be genetically identical (i.e. a single, large clone). It is not improbable that a single, long-lived clone of this kind (and we have no idea how long they may live) may eventually occupy an area of many hectares. I believe that self-sterile clones of considerable size could explain some difficulties in obtaining viable seed from what superficially look like populations of some wetland plants.

For example, a stand of about 12 hectares of jointed twig rush (*Baumea articulata*) I have visited each year for more than a decade has never matured seed properly, and the immature seeds on every stem develop a characteristic twist that I have not seen in any other population. Yet in cultivation, the same plant will set apparently viable seed if it flowers at the same time as other clones of this species from other locations. Even the very abundant and widespread cumbungi (*Typha* species) are commonly found in what appear to be single clone populations. Flowerheads of both native species can vary considerably between wetlands and dams in the same general area, yet are often identical within localised 'populations'. As even the flowerheads of seedlings from a single flowerhead also vary, it is quite likely that many such stands are solitary plants, however large they may be.

If genetic variability is to be retained in revegetation work, it is obviously desirable to bring in planting materials from a number of sources, but how far away should these be at a maximum? In the USA, where considerably more work has been done on revegetation biology and for much longer than in Australia, 80 km is widely accepted as local enough for source materials for planting. Even this figure is regarded as too restrictive by many. Yet it is just a guess and not some scientifically validated figure, as is immediately obvious when it is converted into the distance units used in that country: 50 miles.

All guesstimates of appropriate collecting radii are usually based around general experience with terrestrial plants that are able to spread overland by seed or by stolon. Presumably, a population front advancing overland will tend to select for plants best adapted to the new areas being colonised, and further adaptation will occur over many generations.

Such ideas don't apply as readily to aquatic and wetland plants because of the fragmented and short-lived nature of their habitats. Even the largest lake will silt up and turn into land over millions of years, and the lifespans of many smaller wetlands are measured in thousands of years. Conditions within each wetland change with time and siltation, some plants dying out when conditions are no longer suited to their growth and others moving in at that stage. Many of these wetlands are isolated from others, so plants and seed can't just creep into them from an adjoining swamp or creek. Yet new plants do appear, carried in by birds as seeds, turions or tiny fragments, and as seed blown by the wind.

We know surprisingly little about how most of these species spread, but it is obvious that they must be adapted to 'jumping' from one site to another as in many cases the only alternative route would be the impossible one of growing overland. For some, the adaptations are obvious. Cumbungi are found almost everywhere in Australia because their tiny but copiously produced seeds are attached to fluffy filaments that can carry them great distances in wind.

Close inspection of the seeds of many other wetland plants will show various hooks, bristles and projections designed to catch onto feathers or fur. Some of these seeds are almost impossible to dislodge once they are wedged, especially in down feathers. The viability of such seeds is usually years, while waterbirds moult their feathers at least annually. During that year between moults, some species of waterbirds may travel thousands of kilometres, following rainfall and the resulting filled wetlands anywhere on this continent or even overseas.

Moulted feathers carrying seeds are most likely to fall in a wetland (waterbirds tend to settle in one place for their moult), where they will float until lodging around the water's edge. These are ideal conditions for germination of many aquatic seeds: a dry dormancy followed by soaking and then stranding on waterlogged soil. There is a further advantage to winged travel for these seeds: the feather is nitrogen-rich so it will rot quickly and also release a variety of other nutrients around the growing seedling. This gives the young plant a significant boost as planting identical seeds onto infertile mix, both with and without an accompanying feather, will show.

It is not surprising that many aquatic plants are found over vastly larger areas than most terrestrial species. For example, tall spikerush (*Eleocharis sphacelata*) crops up frequently even on the bare, compacted clay of new dams across most of Australia, New Guinea and New Zealand. Plants from extremes of its range show little difference in appearance, growth and climatic responses, which is not surprising when you consider that the parent plant of the tall spikerush in the dam just up the road could be thousands of kilometres away.

It would make little sense in evolutionary terms for such a widespread plant to develop significant adaptations to purely localised conditions because its primary adaptations must remain suited to spreading by migratory birds (photo 27). This species is widespread because its seeds are designed to be efficiently spread between wetlands, and must be able to grow well whether they have germinated in the tropics or in southern Tasmania. For such far-travelling plants, there is no real reason to insist on using only 'locally' sourced propagating materials.

Tall spikerush is not exceptional among aquatics; indeed, in terms of far flung natural distributions it is somewhere around the middle of the range. There are many other very widespread aquatics, some ranging not only through Australasia but even across all other continents with the exception of Antarctica. Wetland plants found naturally across much of the world, in apparently identical forms, include *Brasenia schreberi, Cyperus polystachyos, Hydrilla verticillata, Isolepis fluitans, Phragmites australis, Potamogeton crispus, P. pectinatus* and *Schoenoplectus pungens.* Others are more restricted to Asia, the Pacific islands and Australasia, including *Baumea articulata, B. juncea, Philydrum lanuginosum, Potamogeton ochreatus, Schoenoplectus validus, Typha orientalis* and *Lepironia articulata,* although the latter is indigenous in Madagascar as well.

Many more comparable examples could be added to this brief list without even mentioning wetland ferns, plants of coastal wetlands or even the many wetland species that are very widespread within Australia itself. It is often pointed out that some of the exceptionally far ranging plants may prove to be a complex of closely related species when they are studied properly over their whole range. Yet even the variation of widespread species complexes is not so great despite the vast areas they may span, or they would not currently be recognised as a single species. This strongly suggests a much greater degree of gene flow through widely scattered populations than would be possible for plants that are not adapted to dispersal across great distances.

Extensive distributions on this scale are not the rule among wetland plants, of course, and there are many lesser degrees of dispersive ability to be found. The less a plant is able to disperse between wetlands, the slower gene flow will be between its populations and the more variation it will show over a smaller range. Some of this variation will certainly be related to local adaptation, although genetic drift due to small founding populations is also likely to be involved. However, all such variation is worth preserving regardless of how it has originated.

There are no clear guidelines to how locally propagating materials of wetland plants should be collected, but estimates should probably be based on the total range of any distinct *form* or *variant* (not just species) within Australia. As variation in genes is likely to increase towards the extremes of the range, it is obvious that estimates should be centred around the area to be planted. I believe that for most wetland plants, a spread of not much more than 10 per cent of total range (that is, 5 per cent in any direction) would preclude the introduction of genes that aren't already present in the area. At the same time, this should allow access to a reasonable number of separate gene pools for all but the rarest and most localised plants.

Applied to widespread species such as tall spikerush or jointed twig-rush, *Baumea articulata*, this suggests an acceptable collecting distance about 400 km in total or 200 km in any direction from the planting site. By contrast, the small water ribbon *Triglochin alcockiae* would need to be collected from within 50 km of any planned planting site, a distance in close agreement with observed variation between populations of this species. For rare or endangered species we may need to stretch these boundaries further.

I am well aware that some readers will disagree vigorously with these estimates, but there are sound reasons for adopting a much more flexible approach to provenance than has been done in the past. Many populations of wetland plants have always been fairly isolated. Their isolation has increased with the draining and destruction of many wetlands they once occupied because the distances between wetlands are much greater now on average. Thus species with limited ability to disperse between wetlands have had their chances of passing genes on elsewhere reduced.

Some must have disappeared without trace from many areas, along with the wetlands in which they once grew. Our records of previous distributions are poor; few botanists had studied wetlands even as recently as 20–30 years ago. Knowledge of what grows in many areas is restricted to the plants there today, often the weediest and toughest survivors to come through two centuries of wetland drainage, grazing and increasing salinity. Unless we start casting our nets wider, we are almost certainly going to be throwing away diversity and replacing it with imitations of the impoverished ecosystems created over the past 200 years.

There are further implications as well, because isolation of populations is believed to be a major factor in the appearance of new species. If we don't make some provision for overcoming lost or reduced gene flow between populations that are further apart than ever before, we are potentially playing God. By rejecting the use of planting materials because they are supposedly from too far away to be adapted to 'local' conditions, we may well be encouraging a gradual drift to localised speciation that could not have begun under less disturbed conditions.

Of course, this assumes that these now isolated populations are large enough and have enough genetic stability to survive for hundreds and thousands of years without the arrival of new genes from other populations. If they don't, we will be killing them off slowly by insisting on their continued unnatural isolation.

• NATIVE OR EXOTIC?

It is not always clear where aquatic plants with wide distributions have originated, particularly if they are found overseas as well as in Australia. Before we try to eradicate unwanted plants in Australian wetlands, it is a good idea to be sure that we really are dealing with introduced species rather than natives that have just had bad publicity. Waterbuttons, *Cotula coronopifolia* (photo 28), have long been regarded as native, but are now increasingly being listed in various floras as an exotic. I will use them to show how debatable the label 'exotic' can sometimes be.

Waterbuttons were widespread around the Port Jackson settlement just after the turn of the 18th Century, when they were first collected by Robert Brown. He included them on a list of 29 introduced and possibly introduced species for the area; they were one of the two species he tagged with a question mark. As exploration of Australia continued over the next half century, waterbuttons were found in wetlands both inland and along most of our southern coasts.

If waterbuttons *had* first been introduced at Port Jackson, they must have been one of the fastest spreading weeds on record, yet they have not been observed to spread into new areas of Australia at any time. In other words, they were already present when each area was explored. By contrast, in comparable areas of the northern hemisphere (where they are definitely introduced) waterbuttons have spread slowly over several centuries.

The earliest collections of waterbuttons were from South Africa, and, as Brown would have seen them there first on his way to Australia, he may have assumed that this is their original home. However, they are also regarded as native to the southern parts of South America. A distribution covering all southern hemisphere continents is nothing novel; for example, the sedge (photo 29) *Isolepis nodosa* (which no one doubts is native) has a very similar range although it is more often found on higher and drier land around coastal and some inland marshes rather than in these marshes.

Other native salt marsh species are more restricted, such as *Samolus repens* and *Selliera radicans* (which are 'only' found from Australia to South America) and *Juncus kraussii* (which is abundant in both Australia and South Africa, although it may be represented by a different subspecies there). If waterbuttons are really native to all these continents — separated by oceanic distances — or even 'just' South Africa and South America, how did they spread?

Waterbuttons will tolerate salinities nearly twice that of the sea, and pieces can float in seawater for at least 4 months. They will even put out roots during that time, so it isn't hard to see how they could survive a journey across an ocean. At least two subfossil *Aepyornis* eggs from Madagascar (the largest of all eggs known) have floated across the Indian Ocean to wash up in Western Australia undamaged. Yet broken-off pieces of waterbuttons are far less delicate and vastly more common, as can be seen along many of our southern coasts after a storm.

The limited evidence above makes it clear that waterbuttons could easily have been established in Australia before the arrival of the first European settlers. However, it *is* also possible that they arrived with sheep brought from South Africa, and then spread much more rapidly in Australia than they are known to have done anywhere else. There is not enough evidence to decide either way, yet floras are traditionally expected to include a clear statement of which plants are native and which have been introduced.

The botanists responsible for these decisions are increasingly declaring indeterminate species exotic, although there are many others who strongly disagree with them. Because they are so often widespread, aquatics are most frequently suspected as introductions. For example, marsh yellow-cress (*Rorippa palustris*) is now 'officially' a non-native, yet the botanist responsible for the *Flora of Australia* account has freely admitted to me that her decision was arbitrary, and she would be just as happy to see it reinstated as a native on the available evidence.

However, this is not obvious from the formal account of this species, because a simple native or exotic decision had to be made, based *on limited evidence that would not be acceptable in a court of law.* It is often suggested that native species 'temporarily' decreed non-native will be reinstated once there is enough evidence to do so, but this is not necessarily true. Unless their indeterminate status is clearly included in written accounts, within a generation it may be forgotten.

Memories are short. Several years ago a Victorian census of plants included *Berula ?erecta* and *Persicaria lapathifolia* as exotics, and I spent two years eradicating them from my nursery as a result. In the most recent edition they have been reinstated as natives, yet no one involved in the publication of either census can remember why they were declared exotic in the first place or why they were reinstated!

How do we deal with such difficulties? Obviously, if a species is exotic it should ideally be controlled or even eradicated in many situations. However, if there is a real chance that it is native, eradication is the last thing we should be doing. The only temporary solution is to describe all plants we can't confidently place under native or exotic as being of indeterminate origin.

This has the advantage that it emphasises species needing detailed study. Genetic techniques are available to estimate how recently Australian populations have separated from others elsewhere. These wouldn't be easy to organise as living material from many sources would be needed because both native and exotic populations may be present here, as is believed to be the case in *Juncus bufonius.*

Yet indeterminate status is not an answer in itself, because it doesn't help with practical aspects such as whether such species should be planted or treated as weeds. Most such plants are widespread and well entrenched, and there is also the problem of what to do with any that eventually prove to be introductions. Funds and energy for wetland management are increasingly limited, while weeds retain their youthful vigour and enthusiasm for life forever. In practical terms, the decision has already been made for us. 'The continued association of exotic and indigenous plants is something that people worldwide will have to learn to accept, live with, and plan for.' (DR Given in *Principles and Practice of Plant Conservation*)

• SELECTION FOR SPECIALISED PURPOSES

WATER TREATMENT

Water treatment is often a stated goal in created wetlands, but the random use of indigenous plants rarely makes a great difference to water quality. This is partly because too few studies have been done on the plants themselves for useful choices to be made in many cases, while those studies that do exist are mostly too vague or limited in scope to be useful. One that sticks in my mind suggests that *Myriophyllum* has little value in water treatment, but does not even specify the species used or how it was planted. Indeed, from the vague information given, the plants used may just have been dropped into the water and not planted at all.

Perhaps the greatest single problem for water treatment is excessive dependence on overseas sources of information, almost entirely from the northern hemisphere. This creates a subtle trap, because most of the genera used there are also widespread in Australia and many of their species are closely related to indigenous ones. These genera have been used overseas because they are readily available and easily propagated in climates that are far colder than any Australian ones, and not necessarily for any other reason.

Yet such plants are mostly dormant in winter, so their nutrient uptake simply stops for a significant part of the year. These plants are certainly not the ones we should be concentrating on in Australia, where many evergreen species with considerable potential continue to be ignored. Worldwide, an estimated 1 per cent of all plants that may be useful in water treatment have been tested or used, and little enough is known about many of these even now. If we limit our selection to those used in cold countries, we are making much the same error as in early timber plantations, where only familiar northern hemisphere species were used. As a result, Australia now imports elite seed of timber eucalypts from overseas, where their potential has long been understood and appreciated.

UTILITY PLANTINGS

Some native aquatic and wetland plants are potentially saleable, perhaps even exportable, although it is likely to be many years before markets will be developed for most of them. *Vallisneria, Ceratopteris* and *Ceratophyllum* are widely used in aquaria, but are probably in adequate supply within Australia and too low in value to be worth exporting. Various native lace plants (*Aponogeton*) have been harvested from the wild on a large scale for the same purpose, but are now belatedly protected. These are readily propagated from seed, are reasonably high in value, and are easy to handle as dormant tubers, but it remains to be seen whether present legislation can be extended to allow commercial cultivation.

Markets for native pond plants are very limited and probably fully catered for, although some indigenous waterlilies have considerable potential as ornamentals for tropical and subtropical areas. Ironically, more work has been done with these in the USA than in Australia, although several species have yet to be cultivated. Other tropicals including *Monochoria* and various *Nymphoides* will also be readily saleable if they become available.

Edible species of aquatic and wetland plants have been neglected in the current 'bush tucker' rush, although several have considerable promise. Water ribbons (tuberous-rooted *Triglochin* species, photo 30) are difficult to harvest from wild populations, but artificial ponds could be designed to overcome this difficulty. Their tubers are numerous, delicious, and can be quite large on selected plants. They are easily raised in large quantities from seed, and there is also the possibility of hybridisation and further selection for large-tubered commercial strains.

Some other species with promise, particularly in warmer areas, include the gelatinous-stemmed *Brasenia* (a delicacy in China and Japan), the large young tubers of *Bolboschoenus* species and perhaps even the tiny but prolific *Wolffia*. Other natives have had their potential pre-empted by overseas strains of the same species, particularly the water chestnut (*Eleocharis dulcis*) and sacred lotus (*Nelumbo nucifera*). Australian water chestnuts are relatively small, although it may be worth looking for larger-tubered varieties with exceptional flavour or production. On the other hand, most native forms of the sacred lotus are so much adapted to the tropics that they won't produce the fat rhizomes that are the most sought after and readily transported edible part.

• PROPAGATION

Plants usually need to be deliberately introduced to most created wetlands and dams or a random assortment of species will appear with time, most of them invasive. The one exception is where a wetland has been restored on the site of an older one that has previously been drained or is being enlarged. In this case, wetter hollows and soils on the site may still hold a good selection of seed and dormant plants that may be able to recolonise the area before unwanted introductions appear. If these wetter areas are going to be submerged deeply in the new wetland, the soils in them will need to be shifted to somewhere around the future water's edge or they may not germinate.

The main disadvantage of this method is that the dormant seeds often include non-native weeds, quite a few species of which grow and set seed very quickly even under less than optimal conditions. These may overwhelm indigenous seedlings when the opportunity arises in the new wetland. Also, the diversity of the hidden seedbank reduces with time as seeds age and die without germination, so that within 25–30 years it is unlikely that much of value will remain.

On most new sites there is no hidden seed bank, and plants will need to be deliberately introduced. This is still often done by transplantation from other wetlands, causing damage to the plants collected and often introducing weeds as well as desirable plants. Although transplants can be a quick way to establish plantings, this is also the most expensive method. Most of the more widespread aquatics can be raised in quantity from seed, and for many species the smaller plants resulting will establish and grow at least as fast as larger transplants of the same type.

Direct seeding is sometimes suggested as an alternative to transplants, but can be grossly wasteful of seed and often doesn't work. It may be practical to use for some species with larger seed that germinates readily, although this must be spread onto a properly prepared seed bed for reasonable germination and establishment. Seed that is slower to commence growth is often very fussy about germinating at all, coming up sparsely and erratically and usually being swamped by other plants before it has a chance to establish.

• NURSERY PROPAGATION

With increasing regulation of wild collection of plants and seed, nursery propagation of wetland plants is coming into its own. This is not much different in overall principles to the propagation of terrestrial species, with plants being grown from seed, spore, cuttings, layers, tubers, dormant winter buds (turions) and divisions including pieces of underground stem (rhizome).

Nearly all plants that can grow both below water (submerged) and above it (emergent) are most readily transplanted in their emergent growth phase. This is because they form firmer, more compact tissues above water and are less likely to be damaged by handling. It is the preferred production method in the aquarium plant industry worldwide, which produces hundreds of millions of plants annually.

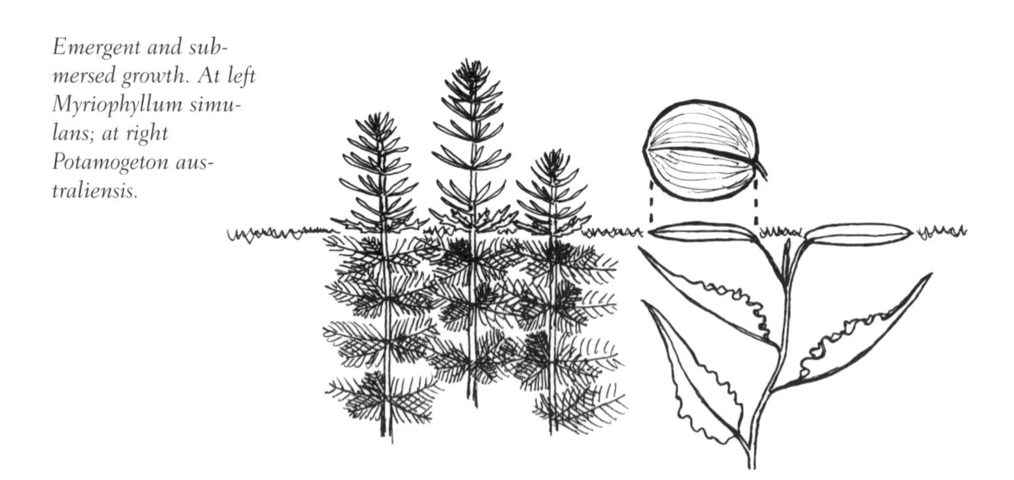

Emergent and submersed growth. At left Myriophyllum simulans; at right Potamogeton australiensis.

If submersed materials are transplanted, they must be kept quite wet at all times during handling before they are planted back under water. Emersed propagation materials can be planted under water, or above it, growing away readily in either case as long as an adequate water supply is kept up to them. Examples of very different looking submersed and emergent growth in the one species can be seen in many *Myriophyllum* species or in *Potamogeton australiensis.* Plants that only produce leaves that float at the water's surface, such as *Ottelia ovalifolia,* should be treated the same as submersed material.

Turions, tubers and other underground storage organs are usually formed as a response to falling water levels or cold, depending on the plant species. They are an easy way to transplant with minimal losses and produce mature plants quickly, but digging them can be a back-breaking job. Make sure that they are planted the same way up as they grew originally or they will waste valuable stored energy reorienting themselves. Some species may not regrow at all if inverted.

Cuttings and layers are easily taken from many sprawling aquatic and water's edge plants, and re-root readily in most cases. Rooting hormones are rarely required for these unless cuttings are taken well out of season, in which case heat will also be necessary to start off a new flush of growth. Massed cuttings of *Myriophyllum* and *Crassula* are sometimes established onto fibrous mats that have been cut into strips. These vegetated strips are then planted out by staking or weighing them down with rocks at an appropriate depth.

However, the plant roots may not bother anchoring themselves through the fibre into the soils below, and as the mat itself breaks down through the action of sunlight, fungi and wave action, chunks can peel away and take the plants with them. The mat itself is unnecessary in any case, as nearly all such species will form a dense tangle of roots when packed together, even without the mat to hold them together. They will do this most rapidly over a layer of rich soil that can be as thin as 1 cm, as it acts primarily as an anchor and nutrient source.

The resulting planting panels can then be cut with any reasonably sharp blade or even scissors. During planting, a little extra soil over the top will hold them down. The stems and foliage should be teased up a little as soil is spread on top of them, so that they will grow up to the light faster. This method gives faster and more permanent bonding between soil and plant than with fibre mats separating the two.

Division basically means splitting a plant through its base or crown. This is another labour intensive method that is generally neither practical nor economical. Quite a few species of wetland plants will be appreciably set back by division, especially if they are divided and lifted from the field in the same operation. Some species, such as *Carex appressa* and *Gahnia sieberiana,* will usually be killed by this treatment. Yet, like many other plants that don't respond well to division, both of these sedges are easily grown if started from seed and transplanted at a smaller size before they have established extensive root systems.

For species that don't set much seed or set it so sparingly over long periods of time that it is not economical to collect it, division remains the main method of nursery propagation. Plants like this are relatively expensive to produce, although their cost is reduced if they are split regularly over a long period. They should be planted out with more care for each individual plant, so that they will mature and self-seed relatively early in the life of a new wetland before other plants have crowded out all potential growing sites.

Rhizome cuttings are taken from sections of the underground stems of plants, and produce new growth from nodes or joints along these stems. In plants that are in active growth there will usually be obvious buds or new shoots at these points. The cut section will generally only need one new shoot on it to re-establish. Dormant rhizomes are best cut just as they start into growth, or the rhizome may rot away from the cut end long before the tips have commenced growth again.

Most wetland ferns can be propagated from spores (photo 34). These are very different from seeds as they have no nutrient store to start them off, and go through a further (tiny) reproductive stage before true ferns are produced. All this takes time, and ferns are easily overwhelmed by faster growing plants if the medium they are grown in has seeds present as well. Sterilisation of the planting medium is essential if large numbers of ferns are to be produced from spores, although small numbers of plantlets will appear around the parent plants spontaneously in some species.

Smaller quantities of many wetland ferns can be raised in finely sieved peat moss wetted thoroughly but not to the point of waterlogging. This should be spread evenly in a clear plastic container with a lid (take-away food containers are perfect), then sealed and kept in bright light out of direct sun. I float the containers in shaded ponds during hotter weather to keep temperatures more uniform; they should be checked every few months. Very small ferns can usually be seen by about 12 months, when they are ready for separating and growing in pots until they are large enough to plant out.

• COLLECTING AND CLEANING SEED

For many wetland plants, seed is the best and fastest means of propagation, but care should be taken to avoid snakes, drowning, and various other hazards while collecting it. Permits are also necessary for the collection of all native plants and seeds on public land, and many species are completely protected. It is essential to check with your state or territory authorities regarding current laws, permits and restrictions. Collecting on private land can only be done with the permission of the owner, but in some states there are further restrictions. Again, check with the relevant authorities before doing anything else.

The most constant hazard in large-scale collecting is that you will inevitably spend many hours out in the sun, so a hat, skin protection and good sunglasses are essential. If you are working near water, remember that ultraviolet light is being reflected up at you as well as shining down. Other ever-present but less obvious hazards are botulism poisoning and tetanus as the bacteria responsible for these are not uncommon in anaerobic mud. These can be lethal, even in tiny doses, but rubber gloves give adequate protection; alternatively, scrub hands and fingernails thoroughly before eating.

Mosquitoes carry a number of the more serious and debilitating diseases spreading throughout Australia at present, so protection against these is recommended but keep in mind that long-term use of some insect repellents will probably be as dangerous to your health as the diseases themselves. Larger leeches can cause a nasty bite and cause bruising that may take months to heal. Some aquatic bugs, particularly back swimmers, have a nasty and long lasting sting, but will only use this if they are handled.

Many snakes, including tiger snakes and copperheads, are common in wetlands because they frequently feed on frogs. Copperheads are often active, even on surprisingly cold days when you would not expect to step on one, so a snakebite kit should be standard equipment at most times of year. Familiarise yourself with their use beforehand, and don't bother to buy any that don't include constrictive bandages to slow the spread of poison through the body. Some cheap, old kits with razor blades are still around. These are almost as dangerous as snakebite and must not be used under any circumstances.

Waders will provide good protection against a bite, especially with heavy clothing underneath, but are extremely hot and uncomfortable in warm weather when snakes are most active. Hip waders can be dangerous; if you slip they will fill and tow you under rapidly, a common cause of death among duck hunters. They should not be used unless there is someone else (without waders) working near you, or if you can be absolutely certain that the area you will be working over is uniformly flat and free of hidden potholes. Thigh waders are shorter, safer and cooler as they can be rolled down when necessary, but these limit your working depth.

On unfamiliar ground I wear a neoprene 8 mm wetsuit long john with boots to match (very stylish, especially with a wet shirt worn over the top). This is reasonably snake proof as well, and also allows you to dip under regularly in hot working conditions. It is sometimes necessary to swim as plants aren't always cooperative about growing where you would prefer them to be, so have swimwear with you at all times. I once swam a river in an apparently isolated area wearing nothing but a large paper bag (for seed) inside a hat, for lack of bathers. After 20 minutes collecting in knee-deep water I discovered three families of campers upstream watching me with alarmed expressions, presumably because of my habit of swimming with secateurs between my teeth.

Some reeds and sedges have very sharp spikes to their tips, and if you are collecting among these you will need more eye protection than sunglasses. Swimming goggles are adequate to protect the eyes themselves, but a diver's face mask provides more complete coverage of sensitive areas. If working among sharp-edged leaves more complete protection is a must, including gloves; even some round-leaved sedges can cause a nasty cut if crushed carelessly in the hand.

The actual collecting equipment can be fairly basic. A pair of secateurs and a large paper bag may be all that is required. Sorting and cleaning is usually best done at a later stage under shade, but don't try it in your tent as it will often produce hundreds of tiny spiders as well as seed. Separate seed from seedheads first, which may require rough handling, especially in the case of sedges. A pair of rubber car mats rubbed together with the seedheads between them will loosen much of the seed, but I find that rubbing by hand gives more thorough results. Wear a glove for this, as many types of seedhead abrade or irritate skin.

After this rough cleaning, the loose rubbish, chaff and seed-eating insects should be separated from the seed. Most commercial operators use a series of fine sieves to grade out larger and smaller detritus, but these are expensive. On field trips I carry only three kitchen strainers of different mesh sizes (about 1 mm, 1.5 mm and 2 mm), which do an adequate job for most types of rough cleaning. These are fairly widely available, although you may need to shop around to find all three grades.

Even after sieving there will usually still be quite a bit of rubbish of a similar size to the seed. This is mostly lighter than the seed and can be separated by winnowing (i.e. putting the mix in a fairly deep bowl and gently tossing it up and down in a breezy place). On a still day you can speed things up by blowing still more gently; this will help even when there is a natural breeze.

The smaller the seed, the slower and more gently you must winnow at the start. With a little practise you will find the right combination of tossing and air movement for each different type of seed. However, if you are careless you can lose all the seed in a matter of seconds. Properly done, winnowing is a very effective process, and it is a delight to watch pure seed emerging from an unappealing mass of chaff, dust and debris.

• PLANTING, TREATING AND STORING SEED

Seed of wetland plants is usually sown on wet soil in a container stood permanently in water. This is referred to as the 'bog method'. The higher the soil surface is above the water level, the more wet soil (with some air and therefore oxygen in it) there will be before seedling roots reach the waterlogged zone. A tray set so that the water is at or just below the soil surface will be waterlogged. This is only suitable for species that can transport oxygen down to their roots as the soil will have little or none in it. These two extremes can be described as 'raised bog' and 'waterlogged bog' methods, although in practice many gradations in between can be used depending on the type of seed to be planted (photo 35).

It is sometimes recommended that a sheet of glass should be placed a few centimetres above the soil to keep humidity high. However, nearly all wetland plants are best sown in full sun, so a glass cover would literally cook them as well as encourage fungal problems. I have yet to see browning off or other signs of damage from dry conditions in seedlings that have been germinated and grown with full light and air movement at all times. By contrast, sowing in shaded conditions results in rather lank, sappy growth, and adds the unnecessary complication of a hardening-off period to what would otherwise be a very simple process. The one use for a cover is to prevent seed from being scattered by rain before it has germinated; however, a sunlit place under a canopy or in a greenhouse will provide all the protection needed.

Seed of submerged plants should be sown in shallow water, usually about 5 cm deep. This should be reasonably warm for good germination, and should be as clear as possible. If the water turns green or brown it should be changed as soon as the seed can no longer be seen from above, replacing it with water of a similar temperature. Submerged plants should only be moved deeper gradually as they grow.

Seed of many (if not most) wetland plants must have light to germinate, so the conventional 'buried alive' method of planting has given quite a few species an undeserved reputation for being difficult to grow. Their response to light makes good biological sense. If they are thoroughly buried under mud or in deep, murky water, there is no point in germinating as it may be a long way to the surface. If disturbances such as floods strip back the covering soil or move them to shallower waters, the seeds will germinate as soon as they detect enough light.

Sedges are a good example of plants whose seeds usually require light for good germination, but some are also reputed to respond to 'secret' pretreatments. These are probably based on the use of growth-stimulating substances called gibberellins, but there are 70 of these known and few of them are commercially available. Most plants respond to one specific gibberellin and the obvious type to try for sedges would be the one used to produce uniform sprouting in barley (during malting) as the grass family is closely related.

Sedge seed that is stripped from the plant before it is completely ripe may also germinate rapidly if kept warm and wet, even in the dark. This presumably forces it to grow as a last ditch attempt at survival, as it isn't mature enough to survive drying out. However, timing of the harvest is critical for this method, and requires some experience. Light-responsive seed nearing the end of its viability will also germinate even in darkness (a dormant seed gradually uses up its supply of stored nutrients, however slowly the embryo metabolises). This also makes sense, as a seed that will shortly have used up all its reserves has nothing to lose by sprouting in the dark. After all, it may find that it is only a short distance from the surface.

If in doubt when sowing wetland seed of any kind, plant it on the surface initially but protect it from sparrows and other seed-eating birds. A sparse layer of soil over seeds may allow enough light through to allow their germination as well as protect them, although this will often slow down the process as well as make it more irregular. Very fine seed should not be covered at all, but should be mixed with very fine grades of sand (e.g. silver sand or hourglass sand) to help spread it out evenly during sowing. This is particularly important if the resulting plants are to be planted out bare-rooted when they are large enough rather than pricked out and grown in pots.

Timing or season can be important in planting. Seed of winter-dormant species often has an in-built dormancy period, and will not germinate until around the same time as mature plants of the same species commence growth. Although such seed can be sown immediately, there is an increased risk of it being overrun by weeds before it begins to grow, so it is often best planted out closer to its peak germination season.

Some wetland seed is very difficult to germinate, or even apparently impossible with the methods presently used. *Baumea* species (photo 32) are the prime example, most of them producing copious amounts of what looks like healthy, viable seed with an apparently perfect embryo inside. Yet germination reports are scarce and contradictory for some species, and no one has had any success at all with the others. Some stands don't produce viable seed, which may reflect a cross-pollination problem (as has been discussed earlier).

However, it may be difficult to obtain much fertile seed from some clones even when they are fertilised with pollen from another source, which suggests that other factors such as fungal infections or adverse growing conditions may be playing a part. Or perhaps this 'problem' is only in the minds of the people who try to grow them. The seed is copiously produced but seedlings are rarely seen even in nature. Perhaps little of it is intended or expected to germinate.

Baumea seeds are relatively large and heavy. They will float but certainly can't be spread by wind. Unlike the seed of some other sedges, there is no sign of barbs to hook onto fur or feathers. Yet most *Baumea* species may be widespread across much of Australia and even beyond, suggesting that they are adapted to travel in the gut of birds. The combination of abrasion and digestive juices in bird guts might well destroy most of them, but it would only take one or two to come out pretreated and ready to grow to keep the species going. The manure they would be deposited in would also speed their growth.

Seeds of some native *Potamogeton* species are known to have survived such treatment by ducks and subsequently germinate readily, although the bulk of them are broken down and digested. Similarly, I have observed occasional *Gahnia sieberiana* seedlings sprouting in crimson rosella droppings that had been tagged after the parrots were seen feeding on the bright red seeds. Most of this seed must also have been destroyed and digested. Perhaps some with particularly hard shells are intended for reproduction, while most are softer-shelled and are mainly produced to attract rosellas or similar birds as dispersal agents.

Is it simply a matter of luck whether any survive? Whatever the cause of unexplained germination problems, it is obvious that to find solutions, or at least explanations, we need to learn more about what would happen to such seeds in nature.

Seed that isn't planted immediately is usually stored dry and cool, as with most terrestrial plants. Dry storage seems to work well for many wetland species, but I have absolutely no doubt that this treatment also shortens the lifespan of many others. This is already well documented for some exotic aquatics, and is also true for at least a few natives. However, there are complications and we are still a long way from understanding viability and germination of quite a few species.

For example, *Ottelia ovalifolia* often grows in temporary ponds, although mature plants don't survive drying out. Seedlings appear not long after their pond refills, so seed must retain viability even under quite dry conditions. Yet in cultivation this seed is often killed quickly by dry storage and is best sown fresh. Perhaps the key here is residual moisture in the pond soils, or the seed may survive better if left in the fruit rather than separated.

The expected viability of many wetland seeds kept out of water is only a few years. Yet we know from wetlands that have been drained for decades and then reflooded that quite a few types of seed will germinate 20 and even 30 years after they were formed, perhaps even more. If you need to store seed with a reputation for short viability, keep it cool and dark in water or moist soil depending on its preferred growing conditions. This works well even for the seed of quite a few tropical waterl ilies (*Nymphaea*), and is also likely to be appropriate for many other aquatic and water's edge plants.

However, dry storage is sometimes the simplest and most effective option, and may even be necessary to trigger germination as it is believed that some types of seed coat are designed to split when wetted after a prolonged dry period. Many *Nymphoides* species respond in this way, with a carpet of seedlings appearing in a newly flooded pond even if the parent plants have been killed by drought. By contrast, the seeds of sacred lotus have a thick seed coat that must be filed through to the embryo to trigger germination or they will lie dormant for decades if left to themselves. This is one seed that is indifferent to dry storage; there is even some evidence that natural germination of the dry seeds may improve after a century or so!

• SOIL, FERTILISERS AND WATER

Soil and water requirements for propagation are much the same as for planting out, which is looked at in more detail in the next section. However, the soils used in a nursery situation may need to be nutrient-rich for faster initial growth, and of course they must be fairly sterile to avoid competition from weed seed. Most nutrient- and lime- or dolomite-free commercial potting mixes make a suitable starting point for wetland seed mixes, or you can use almost any loamy topsoil if you have access to a soil-sterilising kit. Nowhere near as much soil needs to be used for propagation as in the wetland itself, so you can also afford to add more expensive 'ingredients' for better results.

For example, I use up to 10 per cent vermiculite (heat-expanded mica flakes) in all trays because this holds a substantial reservoir of water so that drying out is not as critical on days of extreme heat. Peat should not be used for this purpose as it is very difficult to re-wet once dry; its harvest damages some types of wetlands; and as it is largely imported and not always entirely sterile, it is a potential source of exotic weeds. Blood-and-bone is probably the best all-purpose, slow-release fertiliser and will not damage the roots of less salt-tolerant species as synthetic fertilisers sometimes do. This is added at 0.2–0.5 litres per 100 litres of mix, depending on how heavy a feeder the plant being grown is and how long it will be left in the original mix before repotting.

Few soluble chemical fertilisers are formulated for aquatic use and their component parts are used in unpredictable proportions by wetland plants, so some of the chemicals they contain leach out into water instead of being absorbed. They are also basically a mix of inorganic salts, which quite a few wetland plants don't like, and some stress is probably caused by their use even at lower concentrations. Thus comparisons of plants of the same species grown side-by-side with blood-and-bone versus chemical fertilisers often show different tolerances to insect attack. Aphids in particular can become a problem with the use of synthetic fertilisers, which is not surprising as they are markedly attracted to unnaturally soft and elongated plant tissues.

However, the most obvious effects of salts only appear at increasingly high concentrations that no one would think of applying as fertiliser. Hard or saline waters will have similar effects, reducing flowering and general vigour as salinities reach around 2000–3000 parts per million (ppm). (This older measuring system is often replaced by deciSiemens per metre, but is used here because it is much easier to understand than the more abstract conductivity readings.) By around twice this concentration, most species of aquatics will be dead or very obviously unwell.

Tolerance to salinity varies considerably between aquatic plants (photo 33), although published information on this is scarce at present. In the next chapter, approximate indications of salt tolerance are given for many species. Moderately salt-tolerant species usually appear healthy even at concentrations around 1500 ppm, and perhaps up to 2000 ppm. Very salt-tolerant plants will usually be unaffected at levels around 3000–4000 ppm, although this may vary with their seasonal cycles. As a general rule, salt-tolerant plants will also be tolerant of comparable figures for water hardness, which is a measure of calcium, magnesium and other ions present rather than sodium.

Hard waters will also often be quite alkaline as well (i.e. with a pH well above 7, which is the neutral midpoint from which acidity and alkalinity are measured). The pHof water does not seem to be as important as that of the soil, so plants that won't toerate even slightly alkaline soils will grow perfectly well in a slightly acidic soil even if their water is fairly alkaline. This is another area that needs considerably more study than it has received to date, but as a rough guide most aquatic plants will grow well at soil pH of 5–7 (i.e. moderately acid to neutral).

Many will continue to look well and apparently grow healthily at a pH as low as 4, while some of the more rugged and widespread sedges have been grown successfully in mining wastes at around pH 3. Some species (especially estuarine, inland and salt-tolerant ones) will also grow in alkaline soils. Although the upper limits for these are not known in most cases, few are likely to thrive much above a soil pH of 9. Any plants tolerating at least pH 8 can be described as lime-tolerant.

The soil conditions for plants should be matched reasonably closely to those in the wetland to be planted. If there is likely to be a marked change in salinity, hardness or pH of the water during planting out, plants should be acclimatised to this beforehand. Dormant tubers will not be affected by such changes unless it involves changes to soils and waters outside the limits of their natural tolerances.

• PLANTING OUT

Wetland soils can be very different from terrestrial soils; after all, they have often had centuries or even millennia of the best topsoil and nutrients from the surrounding lands washing in. It is not economical to try to imitate mature wetland soils, and the best we can hope to do is create artificial soils that will provide an adequate substrate and nutrient in the short term until natural processes have taken over to build up something closer to the real thing.

All too often, soil prepared for a created wetland is little more than a blanket of topsoil, usually lacking anything like a balance of the nutrients to which many wetland plants are adapted. Worse still, some sites are simply left as bare clay (often quite compacted), so it is obvious that only a very limited range of plants will be able to grow on them. However, it will probably never be possible to accommodate the needs of all species with an artificial soil, and those with more specific requirements should be added only as wetlands and dams mature.

In a new wetland, all we should aim for is to get the 'framework' species well established. These are the plants that will provide its basic structure and start accumulating nutrients and building up something more like natural soils. Once these framework plantings are established it is usually possible to add fussier plants to round out the ecosystem.

One of the most important qualities for artificial wetland soils can probably best be described as texture: the soil must be reasonably soft and penetrable by plant roots. This is not a problem where a reasonable amount of topsoil has been stockpiled, but if only clays are available these should be broken up to a depth of about 7–10 cm for shallow water and water's edge species. Ideally, this breaking up would be in two parts, an initial run over that will usually leave fairly coarse clods, and a second, more thorough job a few weeks later.

Timing is important here. If broken up months before the wetland fills, the clay will have time to recompact to some degree. The finest, softest and most silt-like conditions can be created with some types of rotary- and tractor-driven hoes after the once-worked clay has already been shallowly flooded. Shallow water and submerged plants particularly benefit from such soft soil textures; the latter will usually be happy with a worked soil depth of 3–5 cm, depending on species.

Worked clays can compact again quite rapidly even while submerged, but the aim of the working is to open them up enough so that plant roots will penetrate right through them quickly. Some clays are nutrient-poor and root growth will be slow in these, allowing the soil to slump back into its earlier condition before the roots improve its structure. The addition of nutrients, organic matter, sand and similar materials may be required to keep such clays more porous for longer or to speed up plant growth.

Cost and availability are the main factors in choosing these. For example, spoiled straw, crushed brown coal and even rock dust from basalt or granite are possible additives and may be inexpensive locally. These should be incorporated with the initial ploughing up of the clay, and even a bulldozer can work them in reasonably thoroughly if the operator is skilled. However, note that the first two of these may decrease pH, while the next two will increase it.

Volcanic clays have most nutrients needed already present, although the addition of nitrogen in the form of blood-and-bone will speed plant establishment. This is thinly spread over the area, then covered with the planting soil to minimise nutrient leaching into the water. Soluble chemical fertilisers should not be used for the same reasons as in nursery propagation. The use of these may explain why some plants sometimes over-run others at an unexpected rate in new plantings, with the more salt-tolerant species having an advantage over those that have been set back by the presence of chemicals to which they are not adapted.

Of course, it is quite possible to leave a wetland in a relatively raw, nutrient-deficient, compacted state and allow natural processes to gradually build up enough soil for a more typical ecology to develop. This will severely limit the species that can be planted initially, and also encourages the seed of colonising species to establish in the absence of competition. As most of these are introduced weeds, this is neither particularly useful nor welcome.

Planting for most wetland and aquatic plants should be done in somewhat shallower water than they are expected to occupy later, as they will spread themselves deeper into their natural zones as they establish (photo 40). Submerged plants may need to be planted much more shallowly initially because the water in a newly developed wetland or dam will often be murky and there may not be enough light for them to grow deeper at first.

Ideally, water levels in a newly planted wetland should be kept close to full for the whole of the first growing season in order to allow plants to grow as quickly as possible under ideal conditions. This can't always be arranged, so planting may need to be staggered to take advantage of the natural cycles of both the plants sand the wetland (photos 36 and 37). For example, autumn-dormant tubers should be planted out long before the wetland begins to fill so that they come up as soon as conditions suit them. On the other hand, submerged plants are probably best left until the second season so that they can be planted deeper (the water should have cleared considerably by this time), where they will be unaffected by anything but significant drought.

Submerged species are best planted bare-rooted. Trying to plant out large numbers of these in pots is slow and will often result in serious drying out, and their soft tissues are easily mangled and broken if they hang over the pot edges. However, most other types of wetland plants are also naturally pre-adapted to bare-rooted planting, partly because they can and do replant and reorient themselves after flood damage and partly because they must be able to routinely survive long flood periods with little light and reduced oxygen levels.

Advantages of bare-rooted handling include reduced freight costs (no extra weight of waterlogged soil to ship, or problems preventing slop!), and greater speed and ease of planting. With tube stock, the planter may have to make repeated trips over long distances dragging heavy boxes of pots in marshy soils unsuited to walking. By contrast, with bare-rooted plants many hundreds or even thousands can be carried in a backpack. This ease of handling in aquatic and semi-aquatic plants explains why, among all the world's important crops, only wetland species such as rice and taro are routinely transplanted as bare-rooted seedlings or divisions.

Unless the texture and materials of the soil used for propagating is closely matched to that of the wetland to be planted, bare-rooted planting may be essential for survival of the plants. For example, *Triglochin lineare* and *T. alcockiae* frequently grow in relatively short-lived pools with a clay substrate. If they are planted out from pots in any of the usual soil mixes (clay mixes are almost impossible to use in pots), a pocket must be dug into the clay for the planting.

As soon as the water level has fallen below the clay's surface, the newly planted seedlings are effectively trapped in a little 'cup' that refills only very slowly from the surrounding clay, if at all. At this stage, the seedling can transpire water much faster than it will seep back into the planting pocket, so it may dry out completely within less than a week in warmer weather. This is long before it would have had a chance to put roots out into the surrounding clay; indeed, there is some evidence that the roots would have trouble penetrating the clay walls of its miniature prison anyway.

By contrast, the same seedling with its roots washed bare could be planted by opening a wedge-shaped hole with a shovel in less than a second. A dab of fertiliser can be dropped in, and the clay pushed firmly back around the plant roots in seconds more. The contact between roots and soil is complete immediately; within another month or two the plantlet will be fully established.

A pack with a thousand bare-rooted seedlings of either species of *Triglochin* would weigh about 2–3 kilograms (much of that wet newspaper), while the same number of seedlings in tubes could require 10 or more trips back and forth to the collection point. These arguments could be extended to many other plants as well. Why lug 100 tubes of a *Bolboschoenus* species through marshy ground to plant them rather later than is ideal? In contrast, the same number of bare tubers would fit comfortably into a spacious pocket. They could be planted into dry soil, commencing growth after moderate rains and establishing earlier as well as forming a closer bond with the soil. The logic, economics and ease of practice are compelling, and bare-rooted planting will increasingly take over from tube stock in the created wetlands of the future.

• VERMIN AND WATERBIRDS

Protection against vermin is just as important in wetlands as in terrestrial plantings, and the most widespread pest is the same in both situations: the rabbit. These will even wade in shallow water, trimming plants down to the surface. Although they can be poisoned to reduce numbers before planting begins, more permanent control may be necessary in the form of fencing.

Standard heavy rabbit mesh is a little over 1 metre high, and allows a little extra for an outward curve at the bottom to discourage digging. It is wired onto a conventional strained fence for support. In sandy soils, the small curve at the bottom is rarely enough to prevent digging beneath it, and an extra panel about 30 cm wide (another standard width) should be laid along the ground overlapping with, and attached to, the bottom of the main panel. Loops of stiff wire can be used to staple the outer edge to the ground.

Young rabbits that are large enough to explore well away from their burrows are still small enough to push through standard rabbit mesh, and it only takes one pair to establish a new colony on the wrong side of the fence. Rabbits will usually avoid open areas until they are too large to get through standard mesh, but if there is vegetation near the fence on both sides then another 30 cm width of finer 2 cm mesh should be run along the bottom of the upright rabbit mesh. Unfortunately, this is not a standard width for smaller meshes, so cutting is involved.

As wetlands are usually intended to be a sanctuary for waterbirds and other animals, it may be useful to extend the rabbit fence to exclude foxes as well. A rabbit fence can be made a bit higher using another standard rabbit mesh width of 1.5 metres, with the top 30 cm forming an overhang coming slightly out from the posts to stop foxes from climbing. A 30 cm overlapping panel at the bottom laid as for rabbit fences on sandy soils will keep foxes from digging through, and is much easier to do than digging in mesh, which is usually recommended. Lighter meshes aren't fox proof, as a hungry one can chew through them.

I have used fences like this to protect ducks and poultry for more than a decade. Although any bird that escapes over a fence is almost invariably taken by foxes the same night, those within have remained safe except on one occasion when bushrats undermined a section of fence. Along sections with softer soil than usual, a double width of the 30 cm mesh may be advisable. As the grass grows through this mesh, it becomes almost invisible and confuses any foxes that try to dig in as they don't seem to think of moving back a distance and trying again.

Carp can be very destructive on submerged plants, partly through direct damage and partly through reducing light levels by digging around in mud. The problems of keeping unwanted fishes out of some types of wetland have already been discussed, but if they can be kept out permanently it will be worth poisoning off any existing fishes. This has usually been done with rotenone, a poison that breaks down to harmless substances within about 2 weeks. It can only be used under permit and is now difficult to obtain. A concentration of 0.5 ppm of rotenone will kill fishes in many cases, but this may need to be increased to 0.75 ppm or even more in harder or more alkaline water.

Other treatments include agricultural lime (not quicklime) applied at about 100 kg per 20 000 litres of water to raise the pH rapidly to about 9 (assuming an initial pH of 6–7). A slow increase may allow the fish a chance to partly adapt to this level, and it may be better to aim closer to a pH of 10 to ensure a complete kill. Swimming pool chlorine or sodium hypochlorite also works at a concentration of 4 ppm and will dissipate to harmless levels within a few days. Both these treatments will also kill other piscine vermin including gambusia and redfin, but will also destroy native fishes and frogs, which is not only undesirable but also illegal in most cases.

The main problem with applying any such treatments is mixing them in, as this must be done as quickly and uniformly as possible. The treatments are easier to apply if the dam or wetland already has a drawdown system built in so that water levels can be lowered, or if water can be partly pumped out. A fire pump, with the hoses used underwater to create strong currents, is a great aid to the mixing.

Finally, mention must be made of waterbirds, which can wreak havoc in a new planting. Even native herbivorous species should be discouraged while plants are establishing themselves, but domestic birds should be kept out absolutely. Geese are particularly destructive and will even pull out newly planted materials, apparently out of curiosity. Ducks will take longer to do the same amount of damage but are more persistent.

5 PLANT INFORMATION LISTS

This list of wetland plants is arranged by genera (plural of genus) followed by the family in which the genus belongs. All genera listed include plants that grow in water for at least a part of the year or around the water's edge at normal (non-flooded) maximum water levels. Although this cut-off may seem arbitrary, if it had been shifted even 15 cm higher the number of genera to be covered would have nearly doubled to include many plants that are only incidentally associated with wetlands.

Even within this scope, many genera described include numerous and diverse species. Particular plants are only mentioned if they need different treatments to the rest of their genus. More detailed species-by-species information on range, habitat and related requirements is available in several books listed under Recommended Reading, and also in a companion volume *Aquatic and Wetland Plants: A Field Guide for Non-tropical Australia.*

To save unnecessary repetition assume that, unless otherwise stated, all plants described are perennial and will grow in full sun, in silty or uncompacted soils with some organic matter and pH 5–7, in a hardness range of 50–1000 ppm, and that seed life under cool dry storage conditions is 2–3 years. Other tolerances for shade, salinity, clay and exceptionally hard or soft waters are mentioned where appropriate.

Seed that needs light for germination is recommended to be sown *on* to soil. Moist soils aren't particularly wet at all, while wet soils will yield some water when squeezed, but there is no exact dividing line between them. However, waterlogged soils are sodden and water will fill any hole pressed into these. As waterlogged soils contain very little oxygen, seedlings and roots of any plants not adapted to such conditions will drown in these.

Exotic plants aren't recommended for wetland plantings, and even Australian species should be regarded as potential weeds if introduced outside their natural range. Use only plants known to occur in the general area or catchment of the wetland or dam to be planted. It has been assumed that any necessary permits will have been obtained before propagating material is collected, including permission from the owners of privately owned wetlands.

Aegiceras Myrsinaceae

See general remarks under mangroves.

Alisma Alismataceae

Water plantain (*A. plantago-aquatica*) can form extensive stands in some areas. The seeds (and to a lesser degree leaves) are eaten by many birds. Plants are handsome and shade tolerant, grow even on clay, and probably have considerable water treatment potential as they respond well to high nutrient levels. Seed has at least a 3 month dormancy period, and will germinate in shallow water but comes up more uniformly on waterlogged mud. It is reputed to lose viability within 12 months according to overseas sources, but is so copiously produced that it can be replaced each year without much effort. Make sure you can tell this apart from the closely related and fast spreading introduced weed *A. lanceolatum,* which has leaves that are much more slender even when mature, and is not so tall.

Alternanthera Amaranthaceae

The lesser joyweed (*A. denticulata*) is a sprawling, shade-tolerant water's edge herb, sometimes very abundant, and a food source for the young of some butterflies and moths. It is easily grown from seed or cuttings by the raised bog method, and will establish even on quite heavy clays.

Amphibolus Cymodoceaeae

The sea nymph (*A. antarctica*) is a common shallow water seagrass, and a significant habitat plant in relatively sheltered bays. It can be propagated by division, preferably in pots of silt so that it can be anchored easily during planting out.

Amphibromus Poaceae

There are quite a few swamp wallaby-grasses, all growing at the edges of fairly ephemeral wetlands where they may be flooded at times and bone dry at others. They are best grown by seed planted on waterlogged soil; this usually needs several months dry before it will germinate but otherwise doesn't present any problems.

Aphelia Centrolepidaceae

These are tiny annuals found in flood-prone areas, where they will readily and prolifically self-seed even on clay if not overgrown by mats of larger plants.

Aponogeton Aponogetonaceae

The native laceplants haven't been grown from seed much, but in my experience aren't any harder to grow than exotic species propagated for the aquarium trade. Some such as Queensland laceplant (*A. elongatus*) have been collected extensively from the wild for aquaria, but are now protected. Other, more tropical species are even more attractive so there is considerable export potential for cultivated plants.

All laceplants are readily grown from fresh seed in shallow, warm water, raising the water level gradually as they grow. Although most species are shade tolerant, intense light is important for strong growth and production of seed, and in some cases a dry period is needed once the developing tubers are large enough to survive this. Tubers of some species have been eaten.

Avicennia Verbenaceae

See general remarks under mangroves.

Azolla Azollaceae

These floating ferns (photo 39) are attractive carpeting plants, but can smother the water's surface in smaller dams and pools. They are readily eaten by ducks and some livestock, and contain nitrogen-fixing blue-green bacteria so they have been used as fertiliser as well. Azollas respond well to fertilisation, and propagate themselves at an alarming rate.

Bacopa Scrophulariaceae
B. monniera is a sprawling, small-leaved herb of shallow water and wet soils. It is shade tolerant and may have potential in water treatment. Propagation is usually by division or cuttings, or seed if you have the patience to harvest it.

Barringtonia Barringtoniaceae
This freshwater mangrove (*B. acutangula*) is a tropical tree forming extensive stands in ephemeral tropical swamps. It is widely grown as a striking red-flowered ornamental, and is easily grown from seed or by cuttings.

Batrachium Ranunculaceae
B. trichophyllum is a buttercup relative with fine underwater leaves. It is presently treated as an exotic in most floras, although there is little enough evidence on which to base this opinion. It is easily propagated by division or cuttings.

Baumea Cyperaceae
Twigrushes are a variable group, mostly growing in shallow waters that are often ephemeral, so they are fairly to very drought tolerant once established. Some (particularly *B. articulata*) respond well to nutrients, and have considerable potential in water treatment. Propagation has generally been by division as seed gives unpredictable results, which may vary dramatically between clones, populations and seasons. This has been discussed in more detail in the previous chapter.

Berula Apiaceae
B. ?erecta is a celery-like plant of shallow waters that tolerates alkaline waters and shade and has some water treatment potential. It is usually propagated by division as it multiplies readily, but seed germinates well on waterlogged soil although it may require a dry dormancy of several months first.

Blechnum Blechnaceae
Some water ferns form extensive colonies in and around seasonal wetlands areas, where they are a significant habitat component. These are mostly shade tolerant and fairly drought tolerant. Propagation from spores is not difficult, but needs to be started about 2 years before plants are required for planting out.

Blyxa Hydrocharitaceae
These are grassy annual plants of clear tropical waters, best transplanted while still small and allowed to self-seed before competing plants establish themselves. Several species are eaten in South-East Asia.

Bolboschoenus Cyperaceae
Club rushes can form extensive stands in shallow ephemeral waters, and are attractive as well as important as habitat. The young tubers are good eating; older ones are woody with an unpleasant sulphurous aftertaste. Plants will grow in shade, on clay soils, have some water treatment potential, and are very drought tolerant. They are propagated from seed that germinates readily when fresh, but will need several months of cold, dry storage followed by warm, waterlogged bog conditions in light if not planted immediately. Dormant tubers are easily transplanted.

Brachyscome Asteraceae
These are small daisies, some species of which grow in flooded areas at the fringes of wetlands. They will mostly grow well on clay, and are easily grown from cuttings or divisions. Seed needs 2–3 months dormancy, and should be planted out on wet soil in autumn as it germinates most readily then.

Brasenia Cabombaceae
Water shield (*B. schreberi*) grows best in fairly deep, warm waters and is winter-dormant in southern Australia. The gelatinous young

stems and leaves are regarded as a delicacy in China and Japan, and it has some water treatment potential. It is usually propagated by division or cuttings, although it can also be grown from seed as for *Nymphaea* if enough of this can be collected to be worth the effort.

Caldesia Alismataceae

Caldesias are annual tropical relatives of *Alisma plantago-aquatica,* which can be grown and treated in much the same way, although propagation is from seed only.

Callitriche Callitrichaceae

These are tiny-leaved creepers of shallow waters and wet soils, most of which are becoming less common as their habitats are increasingly taken over by their introduced relatives *C. stagnalis* and *C. hamulata.* They are propagated by division or stem cuttings only as seed is difficult to collect.

Calystegia Convolvulaceae

C. sepia is found twining through reed and rush stems around wetland fringes, often in areas that flood regularly although not for long. The seed is large and easily gathered, but has a hard coat that should be filed through before planting. Stem cuttings and divisions can be slow to establish.

Carex Cyperaceae

All wetland species of these diverse, grassy sedges can form extensive stands, and many of them are significant shelter and seed-producing plants. Most can be propagated from seed easily by either bog method, and will germinate without any pretreatment or dormant period. One exception is *C. gaudichaudiana,* which doesn't seem to set viable seed reliably. Fortunately, this running species can be divided readily, as can the clump-forming tassel sedge *C. fascicularis.* By contrast, deeper-rooted species such as *C. appressa* resent division and will rarely survive it.

Centella Apiaceae

C. cordifolia (synonym *C. asiatica*) is a creeping plant of wetland fringes with kidney-shaped leaves. It has been used in Asia as a broad-spectrum medicinal plant under such names as Gotu Kola, and in Australia for specific complaints such as arthritis. It is easily propagated by division or from rhizome cuttings. Seed is time consuming to harvest but germinates readily in raised bog conditions.

Centipeda Asteraceae

These are small, sprawling or upright daisies found at the fringes of wetlands, flowering as water levels fall. The two species are powerfully and quite pleasantly aromatic, and have both been used as a treatment for colds and sore eyes. Cuttings will take root easily, and seed germinates freely on wet soils.

Centrolepis Centrolepidaceae

These are tiny, tussock-forming plants that grow on wet to waterlogged soils, and tolerate prolonged flooding. They are easily propagated by division, after which they will self-seed freely, especially on clays where there isn't too much competition for living space.

Ceratophyllum Ceratophyllaceae

The hornwort (*C. demersum*) is a floating plant superficially similar to the submerged leaves of *Myriophyllum.* It can form dense tangles, especially in cooler, moving waters and in the shade of taller reeds and sedges, where it provides shelter and spawning sites for a wide variety of underwater animals including smaller fishes. It responds well to fertiliser, so there is potential for use in water treatment. It is reputed to grow best in slightly alkaline waters, but adequate nutrient levels seem more important in my experience. Propagation is by divisions or pieces, simply floated in water as they don't put out roots.

Ceratopteris Parkeriaceae

These are soft ferns that may take root or float, and grow under water or with foliage above the surface. The fronds are eaten by a wide variety of waterbirds, and used as fairly shade-tolerant aquarium plants. Propagation is usually from the plantlets that form along the edges of mature leaves, but spores sown on wet mud will also grow readily.

Chara Characeae

Stoneworts are curious, branching algae that are frequently mistaken for flowering plants. Their distinctive smell has generated another common name, musk grass. The dense thickets of growth are home to a fair range of invertebrates and a spawning site for many smaller fishes. Some species have potential in water treatment although others may die off suddenly in a mass, which is less than useful for this purpose. Others may actively help to clear fine clay particles from water, although the mechanism for this is not yet understood. Propagation is by division; pieces may be teased apart and then weighed down with a stone or clod of clay. Most are tolerant of lime (many prefer calcium-rich water), moderate levels of salinity and some shade.

Chorizandra Cyperaceae

These are moderately tall sedges from wet to seasonally flooded areas, and are sometimes a dominant vegetation form. They are not difficult to grow on wet soil from seed, which is reputed to have only a short viability (this would make it an exception in the family if it were true). *C. enodis* is widely grown as an ornamental, but is usually propagated by division for this purpose. It will grow on clay, and may have water treatment potential. It is very drought tolerant once established.

Cladium Cyperaceae

C. procerum is a giant sedge found in shallow waters and on wet soils. It is uncommon overall but is likely to be a significant habitat species as it can form extensive stands. It is easily propagated on waterlogged soil, if seed is available. However, some populations don't seem to form seed but reproduce by proliferations, and these may be just a single clone that has spread widely in this way. As the proliferations are large, they can usually be separated into a number of plants.

Colocasia Araceae

Selected forms of taro (*C. esculenta*) are an important food in many tropical countries, and may have been the first cultivated crop. Some botanists believe that taro may have been introduced to Australia, but this seems unlikely on two counts. First, the earliest centre of cultivation was in New Guinea, which suggests that it may have been indigenous there. As New Guinea was linked to Australia by a land bridge a few millennia ago, there is no reason to think that taro could not have spread naturally to northern areas, as have many other plants.

The second reason is more compelling. Wild, native taros grow differently from the selected edible forms (they are exactly like wild forms elsewhere), and most of them can't be made palatable by any known treatment! It is extremely unlikely that anyone would have introduced these deliberately. Wild taros are readily grown from seed planted in waterlogged soil and by division of the runners. By contrast, few cultivated taros produce seed (only infrequently at that), and these almost all produce suckers around a central corm rather than runners.

Cotula Asteraceae

These creeping daisies are found from the water's edge to quite deep waters, depending on species. Waterbuttons

(*C. coronopifolia*) are presently treated as exotic in contemporary floras. This shaky conclusion has been discussed in the previous chapter. If considered a native species, waterbuttons are a significant habitat species, particularly in saline wetlands. All species are readily propagated by seed sown on wet soils or by division.

Craspedia Asteraceae
Upright daisies with globular flowers, these are found around the fringes of wetlands where they may be flooded in the wetter months. Some species have been separated as a closely related genus, *Pycnosorus.* These showy plants can form extensive stands, and are a significant habitat for various invertebrates. All are readily propagated by seed kept wet, not waterlogged, and by division or cuttings.

Crassula Crassulaceae
Swamp stonecrop (*C. helmsii*) is a creeping plant of shallow waters and very wet soils. It forms a dense ground cover that will inhibit the germination of smaller seeds of unwanted species, and is fairly shade tolerant. It is usually propagated by division or cuttings, as seed is time-consuming to harvest.

Cyperus Cyperaceae
A diverse group of sedges found from shallow seasonal waters to much drier environments and often forming extensive stands that are an important habitat for birds and many other animals. The larger species (photo 44) are often adapted to flooding, and have potential in water treatment as they respond well to high nutrient levels. Most are readily propagated by division although they can be slow to recover from root loss. Nearly all are readily propagated by seed, with the exception of *C. lucidus,* which is host to a fungus that destroys the immature seed. Seed is reputed to have a short viability but this is probably incorrect, as all species and provenances of *Cyperus* I have tested for any length of time have retained at least 90 per cent viability even after 3 years in dry storage.

Damasomium Alismataceae
Like a miniature *Alisma,* starfruit (*D. minus*) grows in shallow but ephemeral waters and flowers as water levels recede. This annual plant grows best in the absence of dense competing vegetation. Propagation is by seed only, which is best sown into shallow water.

Diplachne Poaceae
These are attractive annual or short-lived perennial grasses of shallow waters and wet soils, flowering most of the year and spreading readily by seed. They are regarded as a weed in rice crops in southern Australia, but are also potentially valuable grazing and a useful seed source for many birds.

Drosera Droseraceae
The forked sundew (*D. binata*) grows on waterlogged soils and at the fringes of wetlands. More northern populations are many-branched, while in the south a single fork is more usual. Propagation is easy from seed, division, and leaf or root cuttings, all of which should be planted into or on wet soil.

Echinochloa Poaceae
Swamp barnyard grass (*E. telmatophila*) is a tall, annual grass found at the fringes of wetlands, and tolerates prolonged flooding. It is propagated by seed, which needs to be stored dry for several months before planting.

Elatine Elatinaceae
E. gratioloides is a creeping plant of shallow waters and wet soils, tolerating shade and growing in running water. It can form dense mats, and is reputed to be an annual although I suspect that it is perennial in

permanent waters. Propagation is from seed sown on waterlogged soil or by division while it is in active growth.

Eleocharis Cyperaceae

Spikerushes are upright to sprawling sedges that grow from quite deep waters to ephemeral shallows. Most *Eleocharis* species form extensive stands, and are significant habitat for nesting birds as well as reed-dwelling species and smaller animals. One species, Chinese water chestnut (*E. dulcis*), produces tasty tubers; the commercial forms are larger than native ones, and have been selected in China. The rhizomes of tall spikerush (*E. sphacelata*) are supposed to have been eaten, but are not high quality. Some spikerushes grow well with high nutrient levels, and may have moderate water treatment potential. Propagation is easy by division, or from seed sown on a waterlogged bog; some of the smaller species produce proliferations.

Empodisma Restionaceae

Twining roperushes are upright, grassy plants that may scramble through other vegetation around wetlands, usually in seasonally wet soils but also tolerating prolonged flooding. They will grow in shaded situations, on clay, and are moderately drought tolerant once established. They are easily grown from seed on wet soil, but are slow to re-establish if divided.

Epilobium Onagraceae

A number of smaller species of annual willowherbs are found around the fringes of wetlands, but only showy willowherb (*E. pallidiflorum*) is habitually aquatic — it is perennial. This species is usually found scrambling up through taller reeds and rushes in shallow to quite deep waters, and is fairly shade tolerant. All wetland willowherbs are easily propagated by seed kept moist to fairly wet. *E. pallidiflorum* seed should be sown on waterlogged soil, and it can also be grown from cuttings.

Eragrostis Poaceae

Most of the canegrasses are inland species found in depressions that are unlikely to fill except during periods of heavy rain, but *E. infecunda* will grow permanently inundated as well. All species are very drought tolerant when established, will grow on clays, tolerate moderate degrees of salinity, and are fairly tolerant of wave action. They are easily grown from seed, which should be stored dry for about a year for best germination. Division is possible, but plants are slow to re-establish.

Eriocaulon Eriocaulaceae

E. setaceum is a feathery plant of ephemeral tropical waters, with curious button-like flowerheads. It is probably an annual, and is readily propagated from seed (which will germinate even after 7 years stored dry) sown into shallow water. The smaller species in this genus are more widespread across Australia, generally at the fringes of wetlands although some may be flooded in wetter seasons. Propagation for these is by seed sown onto wet to waterlogged soils.

Eryngium Apiaceae

Tanglefoot (*E. vesiculosum*) is a variable, usually spiny-leaved herb of wetland fringes that grows on clay and is fairly drought tolerant. It is easily propagated from seed on wet soil, by division and by root cuttings.

Eucalyptus Myrtaceae

Several eucalypts are found in seasonally flooded areas, the best known of these being the river redgum (*E. camaldulensis*) and coolabah (*E. microtheca*). These widespread and common species are a major habitat type in their own right, growing even on heavy clay soils. They will gradually die if regularly kept flooded for longer than they are adapted. Propagation is by seed only. Some relatively salt-tolerant strains of river redgum are available.

Gahnia Cyperaceae

These exist as medium to large tussocks, often forming extensive stands that provide shelter for a variety of animals, and are also fed on by a variety of insects including the caterpillars of one uncommon butterfly. None of the species divide easily, and many will be killed by this treatment. Propagation from seed has its complications, although all species should be sown moist to wet rather than waterlogged. Seed of some such as *G. melanocarpa* and *G. filum* will germinate as easily as a lawn; seed of *G. clarkei* and *G. sieberiana* needs warmth after approximately 12 months dormancy, while *G. radula* sets very little seed at all.

Gleichenia Gleicheniaceae

Coral ferns (photo 45) are scrambling climbers that like to grow with their roots in very wet soils and their tops in bright light. They will tolerate prolonged flooding, clay soils and some shading. Although they are easily grown from spores and are transplantable while very young, they are almost impossible to divide or transplant when older.

Glossostigma Scrophulariaceae

These are tiny-leaved creepers of ephemeral waters, and flower as water levels fall. All are easily propagated from fresh seed or by division. *G. elatinoides* is used as an aquarium plant in Japan, and will probably be more widely used as an ornamental carpeting plant as it becomes better known elsewhere.

Glyceria Poaceae

Austral sweet grass (*G. australis*) is a major habitat plant, forming extensive stands in seasonal waters up to half a metre deep and tolerating permanent flooding. The seed is eaten by many waterbirds, and plants survive even quite heavy grazing by livestock. Plants will grow on clay, and respond well to high nutrient levels so they may have potential in water treatment. They are easily propagated by division and from seed sown on wet soil. This should be kept dry for several months before planting.

Goodenia Goodeniaceae

Swamp goodenia (*G. humilis*) is a small creeping plant found around the fringes of wetlands. It is usually propagated by division, but is not difficult from seed sown wet if you have the patience to harvest it.

Gratiola Scrophulariaceae

These are sprawling or upright creepers of seasonally wet places, and tolerate permanent flooding as well.

G. peruviana is a shade-tolerant perennial that will grow on clay, in quite dense shade and possibly has water treatment potential. It is easily raised from cuttings or divisions.

G. pubescens is annual and must be grown from seed.

Gunnera Gunneraceae

Found only in Tasmania, *G. cordifolia* grows at the edges of more-or-less permanently wet places and is tolerant of occassional waterlogging. It is a densely carpeting species that multiplies rapidly and is easily propagated by division.

Gymnoschoenus Cyperaceae

Button grass (*G. sphaerocephalus*) forms extensive stands along the fringes of waterways, where it may be flooded at times. In Tasmania, entire plains of this species can be found in some wetter areas, and it is a major type of habitat in its own right. Propagation is usually by division, which is slow and not always reliable, but seed is rarely found in many populations, especially on the mainland.

Halophila Hydrocharitaceae
These are unusual seagrasses with broad, rather rounded leaves. They grow from very shallow water to many metres deep. Rhizome sections take root fairly readily but are brittle and easily damaged, so care is required in handling.

Haloragis Haloragaceae
Swamp raspwort (*H. brownii*) is a sprawling herb of wet places that may flood for long periods of time, and tolerates moderate salinity as well as some shade. It is readily propagated by cuttings or division. Seed should be sown on waterlogged soil, but germination can be poor. Several smaller species grow at the fringes of wetlands and are readily propagated by cuttings, division, or seed sown onto wet soil.

Halosarcia Chenopodiaceae
Many glassworts grow in low-lying, saline soils that may fill with standing water after heavy rains. Seed is slow to germinate and should be kept moist only, not wet at any stage. Cuttings may take but are unreliable.

Hemarthria Poaceae
This is a small grass found in wet places, tolerating waterlogging for some time, moderate degrees of salinity, and some shade. It is easily propagated by division. Seed is time consuming to collect.

Heterozostera Zosteraceae
A seagrass closely related to *Zostera,* it is probably propagated in the same way.

Hydrilla Hydrocharitaceae
Hydrilla is a submerged plant that can be invasive in warm, nutrient-rich situations. It is readily eaten (and kept controlled) by waterbirds, particularly ducks and swans which eat all parts. It is easily propagated from cuttings, or from turions and tubers formed when the plant begins to die down towards winter.

Hydrocharis Hydrocharitaceae
Frogbit (*H. dubia*) is a creeper of shallow waters that also floats out over deeper areas. It can cover extensive areas and is probably a significant underwater habitat species. It is propagated mainly by division, but is probably not difficult to grow from seed sown in warm, shallow water.

Hydrocotyle Apiaceae
These are creeping plants with rounded leaves, and grow from wet soils to shallow waters depending on species. All are easily propagated from seed or by division. The semi-aquatic species will germinate on waterlogged soil, but the others should be kept drier.

Ipomoea Convolvulaceae
This genus includes two native creeping plants of ephemeral, tropical waters. Both are propagated as for *Calystegia.*

Isachne Poaceae
Swamp millet (*I. globosa*) is a small, attractive grass growing in seasonally wet shallows and also more permanent waters. The seed is abundantly produced and is eaten by some waterbirds. It is readily germinated in warm conditions after a long dry storage period, and sown in waterlogged soils. The runners can be divided off without difficulty.

Isolepis Cyperaceae
There are numerous small, tufted sedges in this genus, which is mostly found around the fringes of wetlands and sometimes in shallow waters as well. These smaller species are mostly shade tolerant, and are readily divided or can be grown from seed sown on wet to waterlogged soil. *I. fluitans* is a shallow to quite deep water species that forms extensive floating carpets that take root on mud as water levels fall. The stranded carpets are easily divided. Seed of the much larger-growing *I. nodosa* is easily

collected, unlike that of any of the preceding species, which is fortunate as it resents division. This is not a true wetland species as it prefers the drier fringes, although it will tolerate prolonged flooding.

Isotoma Campanulaceae

Swamp isotome (*I. fluviatilis*) is a tiny-leaved creeper of shallow waters and wet soils, often grown as an ornamental as it flowers prolifically. It is usually propagated by division as seed is not easily gathered in any quantity.

Juncus Juncaceae

The rushes are a large and diverse group that can make up a considerable part of the vegetation of many types of wetland. However, they are somewhat overused for this purpose, particularly some that are really associated with drier habitats rather than wetlands. As many species are more-or-less immune to livestock damage, they have also become much more abundant since European settlement. Using present day abundance of such weedy plants as a guide to reconstruction of natural wetlands is not recommended. Use most *Juncus* species sparingly or not at all, as they will usually invade new wetlands of their own accord.

All species are easily grown from their copiously produced seed, which is reputed to have short viability. However, all of the numerous species I have tested germinate at close to 100 per cent, even after 3 years of dry storage, when sown on moist to wet soils. The most aquatic species, *J. ingens,* has male and female flowers on separate plants. It is the one plant in this genus that doesn't reliably set seed, although this varies between populations and from year to year. If seed can be found, it germinates just as well as that of any other species when sown on wet to waterlogged soils.

J. kraussii and *J. pallidus* are reputed to be difficult to propagate by division. In my own experience the success rate for these is every bit as high as for any other species, provided that divisions are taken while the plants are in active growth and not in the colder months. Most of the more upright species have sharp leaf tips, and eye protection is recommended when collecting seed or divisions of these.

Lemna Lemnaceae

These are small, floating plants that can cover the surface of smaller waterbodies but are unlikely to choke the surface as *Azolla* can do. They are eaten by many waterbirds, and provide protective cover for some frogs. All species multiply themselves readily, and are moderately shade tolerant.

Lepidosperma Cyperaceae

This sedge genus is mainly terrestrial, but pithy sword sedge (*L. longitudinale*) and some relatives grow in seasonally flooded areas. Pithy sword sedge can form dense and extensive stands, and will tolerate prolonged inundation. It resents division, and is slow to re-establish. Seed is very difficult to germinate, even if the embryo inside seems perfect, and I have only had occasional successes with it.

Lepilaena Zannichelliaceae

These are fine, tangled underwater mats, all tolerant of fairly saline conditions. They are easily propagated by division or from cuttings.

Lepironia Cyperaceae

L. articulata is a giant, tussock-forming sedge that forms extensive stands in both permanent and ephemeral wetlands. The rhizomes have been eaten, although harvesting them is labour intensive and is not recommended for propagation purposes. Seed is sparsely produced relative to the size of the plants, but germinates readily when sown on waterlogged soil. Beware of the very sharp tip of the leaf when harvesting.

Leptocarpus Restionaceae
A few species of this genus grow at the edges of wetlands and watercourses, but most will only tolerate sporadic flooding. *L. tenax* is the most aquatic, and is often found in areas where its roots may be submerged for months; it is also fairly salt tolerant. It is not difficult to raise from seed by the raised bog method. Divisions are slow to re-establish, even if large ones are taken. Plants are either male or female, so plantings done from division only should include both sexes for seed production later.

Leptospermum Myrtaceae
Many of these are shrubs of wet places, often tolerating prolonged periods of flooding. The timber is hard and quite durable, and was used by Aboriginal peoples for tools that had to see out a lot of work. Early European settlers used them for jetties and for rudimentary buildings. Propagation is easy from seed sown on wet soil.

Lilaeopsis Apiaceae
This is a small, sprawling and rather tangled plant superficially like some sedges or rushes in appearance. There may be several species lumped under the one species name, *L. polyantha,* at present so use of local provenances is essential in planting. All are readily propagated by division or from seed if enough of this can be gathered.

Limnophila Scrophulariaceae
L. indica is a submerged tropical plant that tolerates some shading. It is usually propagated by division or from cuttings, which take root readily.

Limosella Scrophulariaceae
These are small, creeping plants of shallow ephemeral waters that flower as water levels fall. They are easily propagated by division or from the freely produced seed. *L. australis* appears to shed its seed explosively, and will spread itself rapidly by this method.

Lobelia Campanulaceae
Lobelias are low-growing herbs found around the fringes of wetlands. Some, such as *L. pratioides,* tolerate prolonged flooding; others, such as *L. alata,* are generally found on tussocks and other raised surfaces above the average maximum water table. All are easily raised from seed on wet soil, by cuttings, and by division. They will grow on clay, and in the shade of other plants.

Ludwigia Onagraceae
These are creeping plants of shallow water or herbs found around the water's edge. *L. peploides* will trail long distances out from the shoreline while still attached. All tolerate shade although they may not flower if kept too dark. They respond well to high nutrient levels, have potential for water treatment, and are easily grown from seed or cuttings.

Lycopus Lamiaceae
Austral horehound (*L. australis*) is a tall herb that usually grows on the drier fringes of wetlands, but will also tolerate permanent flooding. It can form dense thickets and is useful as shelter for smaller animals. Propagation is by seed sown onto moist to wet soil, and by division.

Lythrum Lythraceae
Some botanists have believed that there are no native species in this genus, but records of pollen in mud are now known to go back at least 2000 years. Purple loosestrife (*L. salicaria*) is a tall herb with showy flower columns, while *L. hyssopifolia* is a sprawling annual. Both grow in shallow water and on wet soils, and are

readily propagated by seed sown on wet to waterlogged soil. Purple loosestrife cuttings strike readily.

Maidenia Hydrocharitaceae

These are fine-leaved, submerged annuals from tropical areas, propagated by seed sown in shallow to moderately deep but clear waters.

Mangroves Various families

Mangroves are a significant wetland habitat in themselves, becoming increasingly more diverse as both species and communities in the tropics. Their roots are home for a wide range of specialised animals and young fishes, including many that are commercially important. Some have been important as honey flora or for timber, and seeds of some species have been eaten (although appropriate preparation is important). Only five species extend as far south as northern NSW (three of them rare there); only *Aegiceras corniculatum* and *Avicennia marina* extend to around Merimbula, and only *Avicennia* southwards from there.

Aegiceras and *Avicennia* have been most dramatically affected by the spread of what is called civilisation, and are the only species to be propagated on any scale so far. However, the seed biology of many tropical mangroves is similar, and the same methods will probably be suitable for these as well (although little is known about the propagation of non-live-bearing species). Seeds of these two genera begin to germinate while still hanging on the parent, and will take root on any deep, silty mud if they are caught there long enough. If the tidal salinity range is suitable, they will thrive.

Seeds don't thrive when conventionally potted, but will do better in 50 cm or longer sections of 10 cm plastic pipe. This should be cut in halves lengthwise and tied back together before filling with soil so that the seedling can easily be freed later. It is important to set up the nursery area in the right tidal range (e.g. *Aegiceras* prefers to be upstream in somewhat fresher waters than *Avicennia*) or growth will be stunted. The planting hole should be augured, and ties cut before the pipe is pushed into the hole for planting. The roots resent disturbance, and the less they are handled the better, so direct planting of seed is the preferred method wherever this can be arranged.

Marsilea Marsileaceae

Nardoos are creeping ferns with clover-like leaves and are found from shallow, ephemeral waters to deeper, more permanent ones. All are fairly to very drought tolerant once established, will grow on clay, and will tolerate moderate shading. They have some potential in water treatment. The spore cases of several species (formed as water levels fall) are edible, and are usually ground into a kind of flour. Propagation is usually by division, but the dry spore cases split after wetting and the spores start to grow rapidly, so propagation by this method may be viable.

Maundia Juncaginaceae

This is a fleshy, almost iris-leaved plant of shallow, ephemeral waters. It is attractive but uncommon and worthy of special conservation efforts in created and restored wetlands. It is usually propagated by division and by fresh seed sown onto waterlogged soil.

Mazus Scrophulariaceae

M. pumilio is a tiny-leaved creeper of wetland fringes that forms carpets among taller plants. It is widely grown as an ornamental for its showy flowers, and is readily propagated by division or from fresh seed.

Melaleuca Myrtaceae
These are tall shrubs of wetland fringes (photo 41), some of which will grow in shallow to deep waters for prolonged periods of time. However, the majority will be weakened gradually if too wet for too long. The timber of many species is quite hard and was used for building and some types of tools; the papery bark of some species was used as a lining for canoes and roofing. All are readily propagated from seed sown onto moist to wet soil.

Mentha Lamiacaeae
Native mints are low-growing herbs of wetland fringes and river banks rather than true wetland species. They have been used medicinally and to flavour foods, and are mostly fairly shade tolerant. Propagation is by seed sown onto moist to wet soils, from cuttings, or by division.

Mimulus Scrophulariaceae
These are creeping and upright herbs of shallow waters and wet soils, with attractive and showy monkey-faced flowers. Most are easily grown from cuttings and division, or from seed sown onto wet to waterlogged soils. The exception is *M. repens*, which is very salt tolerant and may require the addition of some lime to the soil, although an exact balance is hard to strike.

Monochoria Pontederiaceae
These are creeping to upright herbs of shallow waters (photo 42), with attractive columns of blue or sometimes white flowers. Propagation is easy from division. Seed sown in waterlogged conditions will germinate after a few months, but comes up erratically and should be in full sun or it may be prone to fungal attack.

Muehlenbeckia Polygonaceae
Tangled lignum (*M. florulenta*) forms extensive thickets in seasonally flooded areas, and is a major habitat type in its own right. It is very drought tolerant once established, will grow on clay and may have water treatment potential. Propagation is easy from seed sown in moist to wet conditions. Young growth will take root fairly readily, but note that plants are either male *or* female, so cuttings should be taken from both sexes.

Myriophyllum Haloragaceae
Water milfoils include many important species that provide underwater habitat, particularly in more ephemeral waters although some grow much deeper.. Many respond well to high nutrient levels and have potential in water treatment. Many are tolerant of some shade and also of moderately saline waters. All are readily grown from cuttings although *M. salsugineum* can be very slow to put out roots. Seed will germinate readily in clear, shallow waters and even on waterlogged soils, but is not easy to find in some species and populations.

Najas Najadacaeae
Water nymphs are submerged plants usually found in fairly permanent waters, and are tolerant of moderate to fairly high salt levels. *N. marina* can be quite invasive, forming choked tangles of vegetation, and is best not used where it can spread widely. They are usually propagated by division or by cuttings taken when they are in full growth.

Nelumbo Nelumbonaceae
Sacred lotus (*N. nucifera*, photo 38) is a spectacular, umbrella-leaved plant with large flowers, and is found across Asia into northern Australia. It forms extensive stands in deep to quite shallow tropical waters, although some native clones are surprisingly cold tolerant and are grown as spectacular ornamental plants in cold areas, further south even than Melbourne. Clones that produce dormant tubers in winter can be

propagated by division just as they start to grow in spring; cut any earlier they will just rot away before coming out of dormancy.

However, most of the brick-pink and reddish tropical forms don't produce tubers, and should be propagated by division while in active growth or from seed. The seed has a hard coat that should be filed through to the embryo before planting in shallow, very warm water or it may not bother to germinate in your lifetime. It retains viability for at least 200 years even when stored dry, and there is limited evidence that natural germination may even improve after a century or so! The sacred lotus is a gross feeder and requires considerable amounts of rich, organic matter and nitrogen in the soil; these are usually added as well rotted cow manure and blood-and-bone.

Neopaxia Portulacaceae

Water purslane (*N. australasica*) is a sprawling herb of shallow waters and wet soils, forming extensive lawns in some places. This species responds well to fertilisation and has some potential in water treatment. It will also tolerate moderate shade. It is readily propagated by division and by seed sown onto waterlogged soil.

Nitella Characeae

These unusual algae are loosely related to *Chara* but are generally smaller and finer. The information under that genus applies equally well here.

Nymphaea Nymphaeaceae

All native waterlilies are tropical species, growing in fairly deep but often ephemeral waters and often in extensive stands. They provide shelter, seed and edible leaves for a wide variety of animals, and their stems and tubers have been eaten by humans. Best growth is on silty soils rich in organic matter, with some nitrogen available in the form of blood-and-bone. Seed will germinate readily in warm, clear, shallow water, the young plants being moved deeper as they grow. Tubers also produce offsets that will produce new plants if separated.

Nymphoides Menyanthaceae

Marshworts are creeping herbs of deep to shallow waters (sometimes ephemeral), with floating leaves similar to *Nymphaea* in shape. They can form extensive tangles and are significant as underwater habitat. Most species respond well to fertilisation and probably have some water treatment potential. Propagation is by division and node cuttings of the floating stems. Seed germinates readily if it is dried for some time and then wetted to crack the seed coat.

Nypa Arecaeae

Nypa palms (*N. fruticans*) are an unusual mangrove species that may form extensive groves in tropical areas, although they are not as common in Australia as in parts of Asia. Unlike most mangroves, they will even grow in still freshwater. The branching underground trunks can be divided if large specimens are needed quickly, although casualties can be high. Fresh seed germinates readily but slowly, and needs constant heat for good results.

Oenanthe Apiaceae

Water dropwort (*O. javanica*) is a celery-like herb found in shallow tropical waters in northern Queensland, and also extends as far north as Japan where young tips are commercially harvested and eaten in winter. It is easily propagated by division.

Ondinea Nymphaeaceae

O. purpurea is a curious waterlily relative found only in the Kimberleys, growing in waterholes in the sandy beds of ephemeral creeks. Plants form tubers as water levels fall; these were eaten by aboriginal people. Tubers have been separated to multiply stocks in cultivation, but are sensitive to fungal and bacterial attack so *Ondinea* is best protected and left to look after itself.

Oryza Poaceae
These wild, annual grasses are closely related to edible rice, and have been eaten by humans and many other animals. The potential for commercial exploitation of at least one species is being examined. They form extensive stands in shallow, ephemeral, tropical wetlands, where they can be a major habitat component. Propagation is from seed, which needs dry storage for some months before planting.

Ottelia Hydrocharitaceae
These are attractive plants with floating leaves and large, showy flowers that may open at the surface or self-fertilise below water. They are eaten by many waterbirds and cultivated as ornamentals in warmer areas. Propagation of *O. ovalifolia* is from fresh seed planted in warm, shallow water, but is often reported to be from divisions (based on a misinterpretation of how this species grows). The seed germinates as the seedhead in which it is contained disintegrates, so seedlings often come up in tightly clustered and intertwined groups that are easily teased apart. A close examination of such groups will show that there is no connection between the individual crowns.

Pandanus Pandanaceae
P. aquaticus forms extensive, suckering stands on flood plains in tropical areas, where it is inundated during the wet season. This is a major habitat plant, providing shelter for a wide variety of animals. Water buffalo destroy many of the suckers these days, so this species now often grows as single trunks, which reduces its value as habitat. Fencing buffalo out will allow suckers to regrow, but is not practical except on a relatively small scale. However, the giant diprotodonts (which have only become extinct fairly recently in geological terms) may well have caused the same types of damage, so buffalo may just be their ecological equivalent. This genus is propagated from fresh seed in moist to wet soil or from suckers, although this is labour intensive.

Persicaria Polygonaceae
These are sprawling or upright herbs of shallow, ephemeral waters and seasonally flooded places (photo 46), forming extensive thickets in some places. Most species respond well to high nutrient levels, and have potential in water treatment. They are all fairly shade tolerant. Propagation is easy from fresh seed planted on wet to waterlogged soil, and from cuttings or divisions.

Philydrum Philydraceae
This is a curious plant with spongy leaves (resembling those of an iris) and woolly columns of yellow flowers. It is used in Chinese medicine (I have been unable to decipher the text to find out for what!), tolerant of shade, and responds well to high nutrient levels. Propagation is easy from the prolifically produced seeds sown onto waterlogged soil or into very shallow water. Divisions also strike readily.

Phragmites Poaceae
Reeds form extensive stands that are a major habitat type, although a rather limited one as they crowd all other plants out. They have been used as roof thatching, to make paper, for grazing, and the young shoots are edible. Both native species are also fairly salt tolerant, drought tolerant when established, and survive moderate wave action. They have been used extensively in water treatment overseas, an example that has been followed slavishly in Australia.

In the bitterly cold winter climates of the northern hemisphere, *Phragmites* (and *Typha*) are the only large wetland plants available. However, the common species *P. australis* is dormant for a large part of the year, when it does nothing in the way of absorbing nutrients. It would be more sensible to research the wide variety of fast-growing, nutrient-responsive, *evergreen* plants available in Australia than to continue using plants chosen in very different climates under very different limitations.

Propagation by division is done in spring when growth begins again, producing large plants quickly but with much labour. Seed will grow readily after a few months stored dry, but should be separated from the feathery flowerheads before planting. Alternatively, the mature flowerheads can be laid out over the propagating mix during the winter months, relying on rain to wash out any germination-inhibiting chemicals present. Not all populations of reeds seem to set viable seed, so it is necessary to test this to be sure that this method will work.

If reeds need to be controlled or reduced in numbers, a combination of burning or slashing while water levels are low, followed by grazing of new growth, will set them back considerably. If this is done 2–3 months before plants would normally die down, the stands can be set back dramatically.

Pilularia Marsileaceae

This is a tiny, grass-like fern usually propagated by division, although the remarks on spore propagation under the related *Marsilea* may also apply here.

Pistia Araceae

Water lettuce (*P. Stratioides*) forms floating, leafy rosettes that can create a choking blanket over smaller bodies of water. It is probably only native in the Northern Territory and has been declared noxious in New South Wales. It may have some value in water treatment, and tolerates some shade. Plants reproduce rapidly from offsets.

Polygonum Polygonaceae

P. plebeium is a creeping plant of shallow, ephemeral waters and wetland fringes. It is propagated by seed or from cuttings taken from young plants.

Posidonia Posidoniaceae

This is the most grass-like of the seagrasses, forming extensive meadows underwater. Fresh seed germinates readily. Divisions are usually successful but are slow to re-establish.

Potamogeton Potamogetonaceae

Pondweeds can form extensive submerged stands, mainly in deeper, more permanent waters. Most are fairly salt tolerant, will grow well even with some shading, and have moderate potential for water treatment. Some produce turions or tubers that are eaten by many waterbirds, as are the soft, submerged leaves and the seeds. They can be propagated by division or sometimes from cuttings in spring to early summer, although this is unreliable. Seed can be sown in shallow water and germinates reasonably well, but turions and tubers give the most reliable results in *P. crispus* and *P. pectinatus*.

Pseudoraphis Poaceae

These are creeping grasses of wetland fringes, forming extensive carpets in places. Mud grasses are moderately salt tolerant, and will grow on clay. They are easily propagated from seed sown on wet to waterlogged soil after several months of dry storage, or by division.

Pycnosorus Asteraceae

Closely related and very similar to *Craspedia;* the information under that genus applies equally well here (photo 47).

Ranunculus Ranunculaceae
Some native buttercups grow in shallow waters or at the seasonally flooded edges of wetlands. These can form dense but low carpets, tolerating shade and responding well to high nutrient levels, so they have some potential in water treatment. The aquatic species form useful underwater habitat for many underwater animals as well as for tadpoles and fish spawning. Seed has a 3 month dormancy period after harvest, but germinates readily after that time. Propagation by division is easy, and submerged plants can usually be multiplied from cuttings.

Restio Restionaceae
These are grassy plants forming fairly upright tussocks that are sometimes a significant vegetation form in shallow, ephemeral wetlands. The widely grown and very decorative *R. tetraphyllus* will grow on clay, although it prefers soils with a reasonable amount of organic matter in them. Restios are sometimes propagated from seed on wet soil, but are more often divided. Plants are either male or female, so plantings done from division alone should include both. sexes. Now called *Baloskion*.

Riccia Ricciaceae
These tiny, multi-branched, floating liverworts only reproduce sexually in their stranded stage. They are shade tolerant and easily propagated by division.

Ricciocarpus Ricciaceae
R. natans is a curious, floating liverwort with a black, root-like beard. It is shade tolerant, and sometimes forms dense carpets in smaller water bodies although it is never as blanketing as *Azolla*. It multiplies itself readily, even from a single piece.

Rotala Lythraceae
R. mexicana is a submerged tropical plant of ephemeral wetlands, where it is usually an annual although it may be perennial in more permanent waters. It is grown from seed planted in warm, shallow waters or from cuttings.

Rumex Polygonaceae
Most native docks grow in seasonally wet places rather than in wetlands, but some are also found in areas that may be flooded for weeks or even months. These will grow on clay, tolerate some shade, are fairly drought tolerant and their seeds are eaten by waterbirds. Water dock (*R. bidens*) is more aquatic, growing even in permanent shallows. All species are readily propagated by seed sown on wet soil or from divisions.

Ruppia Potamogetonaceae
Sea tassels are slender-leaved plants that grow in a wide range of salinities, from fairly fresh waters to saltier than the sea. They may be among the few plants present in some deeper salt marshes, and their seed and leaves are probably a major source of food for many passing waterbirds. Propagation is from seed sown in shallow water, although divisions may also grow. However, they can't be grown easily or reliably by either method.

Samolus Primulaceae
These are creeping plants of saline soils, growing in seasonally flooded areas. They are easily grown from cuttings or divisions, and from seed sown onto waterlogged soil. Some lime should be included for *S. repens*.

Sarcocornia Chenopodiaceae
Glassworts are succulent, creeping plants of salt marshes and estuaries, and grow on wet soils that may flood frequently. They often form extensive stands and are certainly a major habitat for many smaller animals and indirectly for the many birds that feed across glasswort flats at higher tides.

Propagation is not easy, and newly planted areas must be protected from disturbance for many months, preferably even for years. Cuttings are slow to establish and not always reliable. Divided plants survive more reliably but take time to recover from surgery.

Schoenoplectus Cyperaceae

Club rushes are medium-sized to tall sedges that grow in moderately deep to shallow ephemeral waters, and often form extensive stands. They are more commonly seen as part of a more diverse patchwork of wetland species, and overplanting will reduce their usefulness as habitat. All native species are moderately salt tolerant, will grow on clay, and the larger species can survive some wave action. They respond well to high nutrient levels, and have considerable potential in water treatment. However, many southern populations are markedly seasonal in their nutrient uptake, and are probably not much more useful for water treatment than *Typha* or *Phragmites* for this reason. Propagation by division is very labour intensive, but seed is copiously produced and most of it will germinate quite readily if planted onto waterlogged soil after several months of dry storage.

Schoenus Cyperaceae

These are small, often creeping sedges of wet places and shallow, ephemeral waters. Most will grow on clays, and will tolerate a fair degree of shading. They are readily propagated from seed sown on wet soil if you have the patience to harvest it, or by division.

Scirpus Cyperaceae

S. polystachyos is a largely sub-alpine sedge similar in appearance and habit to the closely related *Bolboschoenus* species. The general remarks under that genus apply reasonably well here, although nothing is known about the behaviour of fresh seed for this species.

Sedges Cyperaceae

The sedges as a group are such an important component of wetlands that it is worth briefly commenting on their propagation separately. Most species can be divided while in active growth, the main exceptions being tussock-forming plants of drier places, which usually have deep root systems. In most cases, seed is the best method to produce large numbers of plants rapidly, although there are problems with some groups as outlined under the individual entries in this chapter. In general, seed is sown by the bog method — raised for more terrestrial species, waterlogged for the aquatic and semi-aquatic ones.

Seed of most sedges will remain viable for at least 3 years stored dry, but some species have an undeserved reputation for short viability. This is probably because their seed has been buried during planting, and most germinate best with adequate amounts of light. Some sedges produce plantlets at the flowerheads, often instead of seed, but possibly by germination of unreleased seed in some cases. This is called proliferation. The resulting plantlets can be pricked out as in conventional potting, and can even be divided in the process if they are already large enough.

Selliera Goodeniaceae

Selliera is a creeping plant of shallow, ephemeral and saline to very saline waters. It will grow on clay, and tolerates some shade. Propagation is by division.

Sparganium Sparganiaceae

The native burr-reed (*S. subglobosum*) grows in moderately deep to shallow, often ephemeral waters. It can form fairly large stands, and can be a significant habitat plant in warmer areas where it is more abundant. It is usually propagated by division, but fresh seed will also germinate fairly well if sown on waterlogged soil.

Spirodela Lemnaceae

Duckweeds are closely related to *Lemna;* the information under that genus applies equally well here.

Swainsonia Fabaceae

S. procumbens is a sprawling pea relative found in seasonally flooded depressions and at the edge of more ephemeral wetlands, often in heavy clay soils. The seed needs treatment with 30 seconds immersion in near-boiling water, after which it should be soaked in cold water for a few hours for best germination.

Todea Osmundaceae

The king fern (*T. barbara*) is a very large and slow-growing fern found along stream banks and in spring-fed wetlands where there is enough water movement to bring oxygen to the roots. It is shade tolerant and very long lived. Propagation is not difficult from spore, but is very slow.

Triglochin Juncaginaceae

There are several smaller species in this genus, which are mainly found around the moist edges of wetlands, rarely in water. These are all annuals, and little is known about their propagation although they are likely to germinate well from seed sown onto moist soils in autumn. Some prefer relatively saline soils.

T. striatum is a larger and perennial relative of these small plants. There are probably at least two different species under this name, a cylindrical-leafed one that is very salt tolerant and a flat-leaved one that is usually found in fresher waters. Both grow in shallow, sometimes ephemeral waters that can be very stagnant, and will tolerate a fair degree of shading. They are readily propagated by division or from seed sown onto waterlogged soil; this stores reasonably well dried, and will germinate at least 18 months later.

The most conspicuous *Triglochin* species are called water ribbons. These larger, strap-leaved plants with tubers are common in a wide range of fairly deep to shallow, often ephemeral wetlands, or even in moving water in some cases. All of these species were previously lumped under the one name, *T. procerum* (incorrectly *T. procera*), but are now known to be at least eight species with some very distinctive variants that may prove to be further species later.

Regardless of species or form, the group as a whole provides food in the form of leaves, seeds and tubers, as well as shelter both above and below water for a diverse variety of animals. The tubers of most water ribbons are not only edible but high quality, although those of *T. microtuberosum* are too small to be worth harvesting, so there is reasonable potential for food production. Tuber size varies between populations of any one species, and selected forms of different species can be hybridised to give larger crops. However, this is labour intensive as it must be done by hand, and it only works for forms with the same chromosome count.

Most of these plants are fairly tolerant of shade, and some of salt. The eastern variant of the true *T. procerum* even occurs in the upper tidal reaches of estuaries. All are very drought tolerant once established. Many species have potential for water treatment, particularly as their growth mainly depends on adequate water and nutrient levels rather than being purely seasonal, as in some other grassy wetland plants.

Larger plants can be transplanted with some effort as their roots are deeply buried, but resent disturbance and take some time to recover. Propagation is easy from fresh seed sown onto waterlogged soil or into shallow water. Light is essential for good germination. Seed is ready to harvest when it comes off readily in the hand, except in *T. rheophilum,* which may even germinate attached to the flower stalk (an adaptation to the fast moving waters in which it is found). As seed is eaten by many waterbirds it may need to be collected a few days before this stage, but will ripen and germinate if floated for a month or two in clear, shallow water.

Typha Typhaceae

Cumbungi species form extensive stands in deep to shallow, often ephemeral waters (photo 49). They are a major habitat type, but a very limited one, smothering most other types of plant growth. *T. domingensis* is reputed to be less aggressively spreading than *T. orientalis,* but I doubt there is any great difference between them except in cooler areas where *T. orientalis* seems better adapted. It has been suggested that *T. orientalis* is a polyploid offspring of *T. domingensis,* but if this is the case they have been separate for long enough that they have developed distinct variants. In any case there is nothing unusual in new plant species originating through polyploidy.

Both native species are colonising plants adapted to disturbed conditions, and will even grow on the compacted clay of new dams. Dried rhizomes yield a good quality flour, the abundantly produced pollen can be used to make a type of bread, and the young shoots are also edible. Because of their large size they are fairly tolerant of wave action.

Typha species continue to be used on a large scale for water treatment in Australia, mainly because they are one of the two most used wetland genera overseas. However, the discussion of water treatment under *Phragmites* applies equally well to *Typha.* Despite the various problems associated with management of cumbungi, some award-winning wetland projects seem to have used one or the other species extensively. As any other plants present have usually been overrun by the time the wetland has matured, this is a little like handing out vegetation restoration awards for bluegum plantations!

Propagation is usually from seed sown on waterlogged soil. This germinates readily and needs to be pricked out quickly as growth is rapid. In many cases, control is more desirable then planting. This can be done by slashing, which should ideally be done when the flowerheads are maturing just before pollen is shed from the upper head. At this stage much of the energy reserves of the plant are concentrated above soil. Follow-up is by flooding and grazing combined, if possible. Domestic geese are particularly good at destroying the young shoots.

Cumbungi is more readily controlled in mature wetlands where a dense underplanting is already established as the seedlings are too small to force their way up among most other plants effectively. Larger plants can be kept localised if they are surrounded by channels and other waters at least two-thirds of a metre in depth, and preferably twice that. If these are at least 2 metres wide, any rhizomes that try to cross them will easily be seen when they put up young shoots, which pull out easily in deeper waters.

Urtica Urticaceae

The native stinging nettle (*U. incisa*) often grows in areas that are seasonally flooded, and will even grow floating out into permanent water. It will grow in quite deep shade. Propagation is easy by division. Seed sown onto moist to wet soil will also come up readily.

Utricularia Lentibulariaceae

Aquatic bladderworts are floating plants with bladder-like traps used to catch minute underwater animals. They are found among taller plants in quite shaded conditions, but are also tolerant of full sun. Although their preferred habitats are acidic waters, *U. australis* seems unaffected by quite alkaline conditions. Propagation is by division, or from the turions formed when the plants are dying away in winter.

Semi-aquatic *Utricularia* are usually referred to as fairy aprons, and grow at the fringes of wetlands (especially smaller ones) in clay to fairly peaty soils. They flower as water levels fall, and are propagated by seed or division. Some species can be propagated more rapidly by keeping them flooded to around 5 mm depth, as they spread faster in this partly submersed form.

Vallisneria Hydrocharitaceae

Ribbonweeds or eelgrasses are grassy plants of deep to moderately shallow waters, spreading rapidly by fine runners. They are eaten by a wide range of waterbirds, and will grow on clay and in quite deep shade. Most respond well to fertilisation and have potential uses in water treatment. Fruits with viable seeds are only found infrequently in most populations of perennial eelgrasses, so propagation is generally by division. The more tropical, annual species are readily multiplied by seed sown in fairly deep, clear waters.

Villarsia Menyanthaceae

Marsh flowers are creeping or tufted herbs of shallow, ephemeral waters, flowering as water levels fall. Most will grow on clay, tolerate fairly to deeply shaded situations, and are drought tolerant once established. Plants can be divided or propagated from seed as described under *Nymphoides*. Erect marsh flower (*V. exaltata*) has two types of flower, neither of which pollinates its own kind readily, so both types should be included in any planting of this species done by division.

Wolffia Lemnaceae

These are very small duckweeds (photo 48), including the smallest flowering plants in the world. An exotic species (*W. arrhiza*) closely related to the widespread native *W. australiana* is cultivated in Asia as food for humans, and is high in protein and carbohydrates. *Wolffia* is related to *Lemna,* and other information under that genus applies equally well here.

Xyris Xyridaceae

X. operculata is a grassy-looking plant of damplands and sumplands, sometimes growing in waterlogged peaty soils for long periods of time. Plants resent division and rarely recover from it, but fresh seed sown onto a wet mix of half fine sand, half peat with some blood-and-bone as fertiliser will germinate readily and the seedlings will grow quite quickly.

Zostera Zosteraceae

These seagrasses often form extensive underwater meadows. The mature 'flowerheads' break off and release seed that germinates soon after. Plants can also be divided but are slow to re-establish by this method.

• COMMON WETLAND WEEDS

This section only includes a very limited selection of the most widespread or potentially serious weeds, as wetland weeds could fill a volume to themselves (photo 43). Many other minor ones may become worse problems with time, as well as serious potential weeds still being offered through nurseries.

Alternanthera

The introduced alligator weed (*A. philoxeroides*) forms dense drifts that will float out over deeper waters. It is declared noxious for all of Australia, although at present it is mainly found in the Sydney to Newcastle area. Clumps growing in water have been reasonably successfully controlled by a beetle introduced for that purpose, but it will also grow on drier land where it seem less prone to insect damage. The clone in Australia doesn't set seed, so further spread is only likely where pieces have been transplanted by accident. Don't transplant other species from areas where alligator weed is known to occur.

Cyperus Drain sedge

C. eragrostis has been spreading rapidly in south-eastern wetlands and wet soils for the past decade or more, mainly in disturbed areas but appearing even in established wetlands. It is important to keep this species out of new wetland areas wherever possible as it almost impossible to control once entrenched.

Egeria

See comments under *Elodea*.

Eichhornia

Water hyacinth (*Eichhornia crassipes*) is a fast-spreading tropical plant that can form a choking mat over open waters, cutting off oxygen to animals below and encouraging mosquito breeding. Its seeds have been known to continue to germinate as long as 30 years after the last flowering-size plants have been destroyed in an ornamental pond.

Elodea

Canadian pondweed (*E. canadensis*) is a fast-spreading submersed weed that can form dense thickets in nutrient-rich waters. It is almost impossible to eradicate, but is not usually a problem in relatively undisturbed or stable areas as it is eaten readily by many waterbirds. *Egeria densa* is similar, but generally much sparser and less likely to become a problem.

Grasses

Various pasture grasses (e.g. *Glyceria maxima* and *Phalaris aquatica*) have been used in swampy, low-lying areas as grazing. Few are important grazing species and, as there may be palatability problems even with these, one wonders why they were ever introduced when there are native alternatives. Other grasses such as Parramatta grass (*Sporobolus africanus*) have appeared as weeds in the wake of the pasture species. Although most such grasses will invade degraded wetlands and disturbed areas such as drains, they are generally not a problem in established wetlands.

There are many other species with minor claims to being wetland weeds in some situations, but two that were introduced as ornamentals are large enough and invasive enough that they can change even the appearance of a wetland. Pampas grass (*Cortaderia selloana*) self-seeds readily on moist to wet soils. The large tussocks are usually dealt with by spraying, as the spray can be targeted into the crown of individual plants rather than spread more generally. The alternative is burning the crown (not just singeing it!) as it is hard to dig out the roots. If you don't want to tackle existing tussocks of this species immediately, keep flowerheads cut before they can set seed; this only needs doing once a year.

Giant reeds (*Arundo donax*) will grow even in shallow water and can form extensive stands near water, as can be seen in south-west Western Australia. Repeated cutting of young canes while they are still soft will weaken and eventually kill the plant.

Juncus

There are many exotic weeds in this genus. Perhaps the most widespread of these are *J. articulatus* and *J. microcephalus*, both of which colonise disturbed ground in ephemeral wetlands. Although these widespread weeds are regarded as a serious problem in some areas, one Australian textbook gives detailed information on propagating and planting them! Needless to say, ecologically equivalent native species such as *J. fockei* and *J. holoschoenus* are more desirable.

J. acutus is found around watercourses in drier and often fairly saline areas. Its spiny-tipped leaves are dangerous to livestock and probably many other animals as well. I have seen this species deliberately planted in wetlands, apparently mistaken for the native *J. pallidus,* even though it doesn't resemble this closely. It is essential (and simple) to be able to recognise this serious weed, which can destroy much of the value of a wetland and is almost impossible to eradicate.

Ludwigia

L. peruviana is presently confined to the Sydney area, but has the potential to become a serious weed across much of Australia. Outbreaks further afield are expected over time, and should be controlled within 18 months to stop them from setting seed, which it does prolifically. *Ludwigia* forms dense stands that swamp other plants, shedding all leaves in winter so it doesn't even provide adequate shelter for wetland animals for part of the year. Broken pieces will reroot when stranded, and this is also one of the few plants whose seeds will germinate, put out roots, and even form floating islands.

Control of this species is still being studied; mature plants are probably fairly easy to kill by drying out. A series of wetting and drying cycles over 2 years should trigger germination in most seeds, and the resulting seedlings are killed by the following dry phase. For more detail see the paper in "Plants and Processes in Wetlands" on this species, which is listed under "Recommended Reading". A closely related species available through nurseries (*L. longifolia,* sold as *Jussieua*) seems less invasive, but is still potentially a serious weed.

Mimosa

The giant sensitive plant (*M. pigra*) is a tall, superficially wattle-like shrub that is a major weed of some tropical wetlands including Kakadu. It can form dense stands that totally change the nature of a wetland. Control is complicated, and is achieved by various combinations of spraying, cutting out and biological control. Contact your state or territory authorities for help if this weed appears in any wetland for which you are in any way responsible.

Nymphaea

There are numerous introduced waterlily species and hybrids in Australia, of which only *N. mexicana* has turned into a serious weed. This species forms dense, choking blankets in deep to shallow waters, but doesn't appear to set much seed in Australia, so its main spread has been by deliberate introduction. If you must plant an exotic waterlily, choose from the dozens of definitely sterile cultivars available, which offer a far better colour range than this species and also flower more freely. Introduced waterlilies look fine in dams and other obviously artificial wetlands, but will look ridiculous and out of place where a more natural effect is required.

Rorippa Watercress

R. nasturtium-aquaticum is one of the most widespread, abundant and fast-growing of exotic aquatic weeds in southern Australia, particularly in cool, flowing waters. It also produces copious amounts of long-lived seed. Although the plants themselves are soft and usually easily pulled out, they are replaced rapidly by seedlings or tiny broken pieces, which take root easily. There is no known long-term control method.

Sagittaria

Despite their common name, some arrowheads (*Sagittaria*) have spindle-shaped or elongated leaves. Several species (particularly *S. graminea* with Alisma-like leaves, and *S. montevidensis* with unmistakeable arrowhead leaves) are scattered in inland waters of eastern Australia, but are usually only found in disturbed sites such as irrigation channels. Both can be pulled up readily by hand, and a second pulling several weeks later should be enough to remove nearly all the pieces remaining.

Salix

Willows were introduced to control erosion on river banks after the native vegetation that originally held these together was cleared for grazing! They offer little in the way of habitat to indigenous animals, whose lives revolve around the steady, year-round drop of eucalyptus-type leaves into streams and the slow release of nutrients as these decay. The canopy of native trees provides light shade, while larger branches that fall in are shelter for many underwater animals.

By contrast, willows cast a deep shade until they drop their leaves in a mass in autumn. The branches fall as smaller twigs, and like the leaves these break down quickly, so streams surrounded by willows support far less invertebrate life than those among native forest. Larger branches will break off readily and take root downstream. The roots of mature trees creep out into streams, accumulating sediment until they alter stream courses.

Willows aren't of much use even for erosion control as the detritus they build up during floods can increase water pressure so much that the trees are torn out of the ground, taking huge chunks of river bank with them. In dams, large willows can suck up tens of thousands of litres of water during hot weather in addition to the normal evaporation losses that could be expected.

Willows can't just be cut down as they will resprout, nor can their roots be dug out as this will cause serious erosion during even the most mild mannered floods. Control methods have usually involved killing the tree itself with herbicide while leaving the roots intact; suitable indigenous plants are then established as quickly as possible to hold the river bank together before the willow roots decay away.

A few different variations on this have been developed, the most appropriate of which seems to be cutting the tree down, then *immediately* chopping pits downwards into the bark of the stump and filling these with a herbicide. Uncut trees can also be killed this way but, as the roots rot away, the trunk is more likely to be pushed over during floods to cause damage to the river bank; the timber is harder to cut when dry, too. Any tops that are cut green should be stacked so that they dry without putting out new roots.

Even supposedly harmless herbicides such as glyphosphates are now known to destroy frog populations, and probably affect some fishes as well.

Newer formulations are supposed to be much safer and are being touted as completely safe to use in wetlands; the old ones were, too! There is already some doubt as to these claims, particularly as the effects of the newer wetting agents on most native animals are not yet known. Most other herbicides are much more wide-spectrum in the damage they can do, and their use should be avoided if at all possible.

Some recent reports indicate that willows can be killed without using herbicides. Cut willows don't have much energy reserve stored in their stump. If the new suckers that appear are cut once or twice soon after chopping the tree down, and the stump is split and pulped over its surface, the tree will die. If only a few trees are involved, old carpets laid across the cut stumps are often enough to smother the tree without any further action being taken.

Salvinia

Salvinia molesta is a fast-spreading tropical plant that can form a choking mat over open waters, cutting off oxygen to animals below. Reasonable control of *Salvinia* has been achieved with an introduced weevil, so if this species is thriving in your area it may be worth introducing any badly chewed-looking plants found elsewhere in the hope that the weevil is present on them.

Typha Cat-tail

Typha latifolia is in the same family as the native cumbungi, but is easily recognised by the dark, chocolate-brown, poker-like flowerheads; the native species have much paler, tan flowerheads. It is a widespread and freely self-seeding weed in southern Australia, displacing or growing with cumbungi in many places. Most sedges and *Typha* are significantly weakened by slashing when their flowerheads are maturing as most of their reserves are concentrated in the top of the plant at this time, while the seeds are still far from ripe. Following up with flooding may be enough to dramatically reduce the numbers of many species. Grazing the new growth will weaken them still further at this stage.

GLOSSARY

artesian waters underground waters brought up from some depth, used in drier areas as a source of stock and irrigation water, but usually very hard.

bywash *see* overflow.

catchment the land surface from which the water to a dam or wetland is collected as runoff.

clone genetically identical plants, whether one very large stand spread over many hectares or many smaller plants obtained by division or cuttings from a single parent.

corm an underground storage organ developed from the stem.

crown the growing centre of a plant; there may be many separate crowns which can be divided in some species.

cutting a section of stem able to grow new roots.

dampland An area where the water table rises to very close to the soil surface during some times of the year.

division a cut splitting a plant into smaller, already complete plantlets.

drawdown lowering the water level in a wetland, sometimes used as a management method for controlling weeds or vermin.

emergent the parts of a plant growing above water.

ephemeral in wetlands, a body of water that regularly dries out for a part of the year.

estuary the section near a river or creek mouth where seawater and freshwater mix.

exotic introduced from another area, usually another country in the sense used here.

genetic drift an inbreeding effect that happens when a very small number of plants (or even just one) colonise a wetland and reproduce to form a closely related population that may be somewhat different (drift away from) the norm for that species.

germination in seeds, sprouting.

ground water underground water (*see* also water table).

hardness a measure of the quantity of calcium, magnesium and other salts present in water, other than sodium salts (*see* also salinity).

hydrologist a person who studies the ways in which water moves through wetlands and underground.

indigenous a plant native to an area (i.e. one that has not been introduced from elsewhere).

levee an earth bank used to divert waters in new directions or hold back flood waters.

mangroves trees adapted to living in areas where water levels rise and fall with sea tides, and salinity can vary from almost pure seawater to freshwater at different times.

node a swelling or break in a stem.

overflow a section of embankment that is made a little lower than the rest, where excess runoff water leaves the dam rather than just over any random part of the wall. Sometimes called a bywash.

peat plant matter that has only partly decomposed because of the absence of nitrogen and some other nutrients; peat is very absorbent so it acts as a water store in swampy soils.

pH a scale used to compare acidity or alkalinity, centred around a neutral point of 7. pH readings below 7 are increasingly acidic; values above 7 are increasingly alkaline.

ppm abbreviation for parts per million.

pricking out separating small seedlings (or proliferations) that have been growing very close together to save space during the germination stage, and planting them out individually.

proliferation when plantlets form in the seedhead while still attached to the parent plant; these can be detached and planted as for seedlings.

rhizome an underground stem.

runner a running rhizome, or sometimes a surface astem producing new plants.

salinity a measure of the quantity of sodium salts present in water (*see* also hardness).

soil seedbank living but dormant seed in the soil of a wetland

spore a tiny seedlike structure containing no food reserves. Produced by ferns and other less advanced plants.

stolon a runner with nodes along it.

substrate any soil or solid artificial mixture in which plants are grown.

sucker a new shoot or growth coming up from the base of a plant.

sumpland a dampland where the water table rises up to and sometimes a little above the soil surface at some times of the year.

terrestrial growing or living on land, rather than in water or very wet places.

tuber an underground storage organ formed on a root.

turion a resting bud, usually formed in winter, from which new plants grow.

water table the upper surface of underground water, which may rise or fall with the seasons. This can be located by digging down to the point where a hole fills with water to a certain level — the water 'table'.

RECOMMENDED READING

J Bradley, 1988. *Bringing Back the Bush*. Sydney, Lansdowne Press. This describes the Bradley Method of tackling weeds in bushland, which can also be applied in wetlands. The section on reclaiming river banks while minimising erosion is particularly useful.

M Brock, 1997. *Are There Seeds in Your Wetland? Assessing Wetland Vegetation*. Many wetlands that have been significantly altered or heavily grazed still have a considerable reserve of living seed in the soil. This booklet describes how you can assess what remains before considering importing plants or seed from elsewhere. Available from Land and Water Resources Research and Development Corporation, GPO Box 2182, Canberra, ACT 2601.

MA Brock, PI Boon & A Grant (Eds), 1994. *Plants and Processes in Wetlands*. Canberra, CSIRO Australia. A collection of papers looking at various aspects of the biology of plants in wetlands, from effects of water levels and grazing to control of some weeds. Fairly technical.

JM Chambers, NL Fletcher & AJ McComb, 1995. *A Guide to Emergent Wetland Plants of South-western Australia*. Perth, Murdoch University. Very incomplete: only sedges, reeds, one restiad and *Typha* are covered despite the title, a total of 14 species of which three are introduced. This book does provide propagation information for many of the species described. However, it is hard to understand why detailed propagation instructions have been included for the three exotic species, particularly as these have no special merits of any kind and are already very successful weeds across much of Australia without deliberate planting.

NC Deno, 1993. *Seed Germination: Theory and Practise* [sic] (2nd edn). Pennsylvania, self-published. This is perhaps the most ambitious practical study of seed germination ever attempted, summarising results of replicated germination trials under diverse conditions for thousands of species. Most of these aren't Australian, although the extensive theoretical summaries are universally applicable. It is available from the author at 139 Lenor Drive, State College, PA 16801, USA.

DA Falk & KE Holsinger (Eds), 1991. *Genetics and Conservation of Rare Plants*. New York, Oxford University Press. This looks particularly at the problems associated with small populations, including the conservation of variation in disjunct and clinal populations.

PL Fiedler & SK Jain (Eds), 1992. *Conservation Biology: The Theory and Practice of Nature Conservation, Preservation and Management*. New York, Chapman and Hall. Useful discussions of causes of rarity in plants, and also loss of biodiversity in aquatic systems.

DR Given, 1994. *Principles and Practice of Plant Conservation*. Oregon USA, Timber Press. Absolutely the best text in this area, and also the most current one.

DA Hammer (Ed), 1989. *Constructed Wetlands for Wastewater Treatment: Municipal, Industrial and Agricultural*. Chelsea, Michigan, Lewis Publishers. A little dated now, but includes some 'historical' material that puts newer work into perspective, and also quite a bit of background information of relevance today.

C Harty, 1997. *Mangroves in New South Wales and Victoria*. Melbourne, Vista Publications. A useful practical guide, including management and replanting.

DL Jones, 1987 (reprinted 1993). *Encyclopaedia of Ferns*. Melbourne, Lothian Books. Good coverage of propagation from spores, including for more difficult species.

WR Jordan, ME Gilpin & JD Aber (Eds), 1987. *Restoration Ecology: A Synthetic Approach to Ecological Research*. Cambridge, Cambridge University Press. This does not provide much on wetlands specifically but is worth reading for the experimental approach to reconstruction of damaged ecologies, and emphasises the unique types of information that such studies can produce.

JA Kusler & ME Kentula (Eds), 1990. *Wetland Creation and Restoration: The Status of the Science*. Washington, USA, Island Press. A hefty volume giving a useful overview of practical wetland work in the USA up to the late 1980s, this emphasises restoration or recreation of natural systems and *evaluation* of the results, an area that is largely neglected in Australia.

AJ McComb & PS Lake, 1990. *Australian Wetlands*. Sydney, Angus and Robertson. This is a useful and readable general account of wetlands with some very nice photography.

N Mackay & D Eastburn (Eds), 1990. *The Murray*. Canberra, Murray-Darling Basin Commission. A broad overview of our largest river system, with specialist contributions covering everything from the physical environment to biology. Reasonably light but informative reading.

NSW Department of Agriculture has produced a number of papers and pamphlets on willows (identification and destruction). These are *Willow Identification for River Management in Australia*, *Willow Control* and *Willows Spreading by Seed*. They are available from NSW Department of Agriculture, Locked Bag 21, Orange, NSW 2800.

NF Payne, 1992. *Techniques for Wildlife Habitat Management of Wetlands*. New York, McGraw-Hill. The USA is far ahead of us in wildlife management, and this book provides a good summary and overview of what has been done there.

N Romanowski, 1994. *Farming in Ponds and Dams: An Introduction to Freshwater Aquaculture in Australia*. Melbourne, Lothian Books. This provides detailed information on setting up various types of siphon systems and pipes with baffles along with dam construction, including alternatives for clay-poor soils and problem situations.

GR Sainty & SWL Jacobs, 1981. *Waterplants of New South Wales*. Sydney, Water Resources Commission. This is still a good general guide, especially for the colour photography, although the names for many species have been changed over the years.

GR. Sainty & SWL Jacobs, 1994. *Waterplants in Australia: A Field Guide* (3rd edn). Sydney, Sainty and Associates. A general guide with about 130 plants from all around Australia, of which one-third are introduced weeds. It also provides good coverage of blue-green bacterial problems.

CM Schonewald-Cox, SM Chambers, B MacBryde & WL Thomas (Eds), 1983. *Genetics and Conservation*. Menlo Park California, Benjamin/Cummings Publishing. This contains various specialised articles looking at variation with particular reference to inbreeding in smaller populations, and preservation of biodiversity.

S Usback & R James (compilers), 1996. *A Directory of Important Wetlands in Australia* (2nd edn). Canberra, Australian Nature Conservancy Agency. This details hundreds of the most important natural wetlands in Australia, state by state.

BIBLIOGRAPHY

Anon, 1995. National Conference on Wetlands for Water Quality Control, Townsville, James Cook University (various sponsors and organisers; no editor cited).

P Adam, 1990. *Saltmarsh Ecology*. Cambridge, Cambridge University Press.

HI Aston, 1973. *Aquatic Plants of Australia*. Melbourne, Melbourne University Press.

HI Aston, 1995. A revision of the tuberous-rooted species of *Triglochin* in Australia. *Muelleria* 8(3), 331–364.

J Bradley, 1988. *Bringing Back the Bush*. Sydney, Lansdowne Press.

R Braithwaite, 1994. Pandanus: Then and now. *Australian Natural History* 24(11), 24–31.

M Brock, 1997. *Are There Seeds in Your Wetland? Assessing Wetland Vegetation*. Canberra, LWRRDC.

MA. Brock, PI Boon & A Grant (Eds), 1994. *Plants and Processes in Wetlands*. Canberra, CSIRO Australia.

VJ Burke & JW Gibbons, 1995. Terrestrial buffer zones and wetland conservation: A case study of freshwater turtles in a Carolina Bay. *Conservation Biology* 9(6), 1365–1369.

J Cann, 1978. *Tortoises of Australia*. Sydney, Angus and Robertson.

JM Chambers, NL Fletcher & AJ McComb, 1995. *A Guide to Emergent Wetland Plants of South-western Australia*. Perth, Murdoch University.

K Cremer, C Van Kraayenoord, N Parker & S Streatfield, 1995. Willows spreading by seed — Implications for Australian river management. *Australian Journal of Soil and Water Conservation* 8(4), 18–27.

NC Deno, 1993. *Seed Germination: Theory and Practise* [sic] (2nd edn). Pennsylvania, self-published.

NC Deno, 1996. *Seed Germination: Theory and Practise (First Supplement)*. Pennsylvania, self-published.

WR Elliot & DL Jones, 1980 to 1993. *Encyclopaedia of Australian Plants Suitable for Cultivation* (Volumes 1–6). Melbourne, Lothian.

DA Falk & KE Holsinger (Eds), 1991. *Genetics and Conservation of Rare Plants*. New York, Oxford University Press.

PL Fiedler & SK Jain (Eds), 1992. *Conservation Biology: The Theory and Practice of Nature Conservation, Preservation and Management*. New York, Chapman and Hall.

RH Froend, RCC Farrell, CF Wilkins, CC Wilson & AJ McComb, 1993. *Wetlands of the Swan Coastal Plain. Volume 4: The Effect of Altered Water Regimes on Wetland Plants*. Perth, Water Authority of WA & WA Department of Environmental Protection.

DR Given, 1994. *Principles and Practice of Plant Conservation*. Oregon USA, Timber Press.

SA Halse, GB Pearson & S Patrick, 1993. *Vegetation of Depth-Gauged Wetlands in Nature Reserves of Western Australia*. Technical Report No 30. Perth, Department of Conservation and Land Management.

DA Hammer (Ed), 1989. *Constructed Wetlands for Wastewater Treatment: Municipal, Industrial and Agricultural*. Chelsea, Michigan, Lewis Publishers.

C Harty, 1997. *Mangroves in New South Wales and Victoria*. Melbourne, Vista Publications.

JH Hawking, 1994. *A Preliminary Guide to Keys and Zoological Information to Identify Invertebrates from Australian Freshwaters*. Identification Guide No 2. Albury, Co-operative Research Centre for Freshwater Ecology.

JM Hero, M Littlejohn & G Marantelli, 1991. *Frogwatch Field Guide to Victorian Frogs*. Melbourne, Department of Conservation and Natural Resources.

P Horwitz, 1995. *A Preliminary Key to the Species of Decapoda Found in Australian Inland Waters*. Identification Guide No 5. Albury, Co-operative Research Centre for Freshwater Ecology.

P Hutchings & P Saenger, 1987. *Ecology of Mangroves*. Brisbane, University of Queensland Press.

P Johnson & P Brooke, 1989. *Wetland Plants in New Zealand*. Wellington, DSIR Publishing.

DL Jones, 1987 (reprinted 1993). *Encyclopaedia of Ferns*. Melbourne, Lothian Books.

WR Jordan, MF. Gilpin & JD Aber (Eds), 1987. *Restoration Ecology: A Synthetic Approach to Ecological Research*. Cambridge, Cambridge University Press.

JA Kusler & ME Kentula (Eds), 1990. *Wetland Creation and Restoration: The Status of the Science*. Washington, USA, Island Press.

GJ Leach & PL Osborne, 1985. *Freshwater Plants of Papua New Guinea*. National Capital District, University of Papua New Guinea Press.

AJ McComb & PS Lake, 1990. *Australian Wetlands*. Sydney, Angus and Robertson.

AJ McComb & PS Lake (Eds), 1988. *The Conservation of Australian Wetlands*. Sydney, Surrey Beatty and Sons.

R McDowall (Ed), 1996. *Freshwater Fishes of South-eastern Australia*. Sydney, Reed.

N Mackay & D Eastburn (Eds), 1990. *The Murray*. Canberra, Murray-Darling Basin Commission.

KD Nelson, 1985. *Design and Construction of Small Earth Dams*. Melbourne, Inkata Press.

R Ornduff, 1996. The breeding system of *Villarsia exaltata*, a distylous species. *Telopea* 6(4), 805–811.

WT Parsons & EG Cuthbertson, 1992. *Noxious Weeds of Australia*. Melbourne, Inkata Press.

NF Payne, 1992. *Techniques for Wildlife Habitat Management of Wetlands*. New York, McGraw-Hill.

M Ralph, 1994. *Germination of Local Native Plant Seed*. Melbourne, Bushland Horticulture.

JH Ross, 1996. *A Census of the Vascular Plants of Victoria* (5th edn). Melbourne, Royal Botanic Gardens.

GR Sainty & SWL Jacobs, 1981. *Waterplants of New South Wales*. Sydney, Water Resources Commission.

GR Sainty & SWL Jacobs, 1994. *Waterplants in Australia: A Field Guide* (3rd edn). Sydney, Sainty and Associates.

CM Schonewald-Cox, SM Chambers, B MacBryde & WL Thomas (Eds), 1983. *Genetics and Conservation*. Menlo Park California, Benjamin/Cummings Publishing.

CA Semeniuk, 1987. Wetlands of the Darling system — A geomorphic approach to habitat classification. *Journal of the Royal Society of Western Australia* 69(3), 95–112.

RJ Shiel, 1995. *A Guide to Identification of Rotifers, Cladocerans and Copepods from Australian Inland Waters*. Identification Guide No 3. Albury, Co-operative Research Centre for Freshwater Ecology.

MB Thompson, 1983. Populations of the Murray River tortoise: The effect of egg predation by the red fox. *Australian Wildlife Resources* 10, 363–371.

MJ Tyler, LA Smith & RE Johnstone, 1994. *Frogs of Western Australia*. Perth, Western Australian Museum.

WD Williams, 1980. *Australian Freshwater Life* (2nd edn). Melbourne, MacMillan.

HBS Womersley, 1984. *The Marine Benthic Flora of Southern Australia, Volume 1*. Adelaide, SA Government Printer.

PA Yeomans, 1981. *Water for Every Farm Using the Keyline Plan*. Katoomba NSW, Second Back Row Press.

W Zeidler & WF Ponder (Eds), 1989. *Natural History of Dalhousie Springs*. Adelaide, South Australian Museum.

INDEX